Paths to a Green World

Paths to a Green World

The Political Economy of the Global Environment

Jennifer Clapp and Peter Dauvergne

The MIT Press
Cambridge, Massachusetts
London, England

MIT Press books may be purchased at special quantity discounts for business or sales promotional use. For information, please e-mail special_sales@mit-press.mit.edu or write to Special Sales Department, The MIT Press, 5 Cambridge Center, Cambridge, MA 02142.

This book was set in Sabon by SNP Best-set Typesetter Ltd., Hong Kong. Printed and bound in the United States of America.

Library of Congress Cataloging-in-Publication Data

Clapp, Jennifer, 1963–
  Paths to a green world : the political economy of the global environment / Jennifer Clapp and Peter Dauvergne.
      p. cm.
  ISBN 978-0-262-03329-9 (hc.:alk. paper); 978-0-262-53271-6 (pbk.:alk. paper)
  Includes bibliographical references and index.
  1. Environmental economics. 2. Environmental policy. 3. Global environmental change. 4. Globalization—Economic aspects. I. Dauvergne, Peter. II. Title.

HC79.E5C557   2005
333.7—dc22

                                                                2004059256

10   9   8   7   6   5   4

For our families

Contents

Illustrations

Preface

Paths to a Green World is a highly ambitious book. It is the first to concentrate exclusively on the political economy of the global environment, striving to integrate the debates within the "real world" of global policy and the "academic world" of theory. It moves well beyond the traditional academic focus on international agreements and institutions in an effort to capture the views on politics, economics, and the environment within the halls of global conferences, on the streets during antiglobalization protests, and in the boardrooms of international agencies, nongovernmental organizations, and industry associations. In doing so, it investigates the debates over globalization, environmentalism, economic growth, poverty, consumption, trade, corporate investment, and international finance. It does so from a variety of angles—economic, political, ecological, and social.

The book does not advocate for a particular perspective on how politics and economics relate to the health of the global environment. Instead, it offers an original typology of worldviews to classify the various debates. This typology is, we believe, parsimonious enough for readers to grasp the key threads with ease, yet nuanced enough to rouse vigorous debate. The book fills, in our view, a critical gap in the literature on global environmental change. It meets an immediate need in the field of global environmental politics, by providing comprehensive coverage of the political economy of the global environment. The typology we propose in the book, we hope, will also meet a much more imposing need: to help scholars, bureaucrats, industrialists, and activists communicate in a common language. This latter goal is perhaps too ambitious, perhaps even naive. But striving to facilitate such dialogue is, in our admittedly immodest vision, worth the risk of seeming arrogant about our ability to traverse disciplinary boundaries.

We have tried our best to explain the complexities of the political economy of global environmental change without disciplinary jargon. Naturally, the book uses terminology; otherwise, it could only skim the surface of the core debates. Yet, at every turn, we strive to explain debates and define terms in ways that transcend disciplines. Our hope is that those from a range of educational backgrounds—including development studies, economics, environmental studies, geography, human ecology, international law, philosophy, political science, and sociology—can use this book for a big-picture snapshot of the core debates.

Paths to a Green World would also function well as a university text-book to introduce the debates on the interface between political economy and global environmental change. Instructors using this as a textbook may want to add case studies of particular global environmental problems. In our own teaching, we add, for example, lectures and readings on the political economy of climate change, deforestation, food security, nonrenewable resource extraction, ozone depletion, persistent organic pollutants, and trade in hazardous waste. But other global environmental issues—like acid rain, biodiversity loss, desertification, energy use, overfishing, genetically modified organisms, trade in endangered species, transboundary pollution, whaling, as well as many others—would work equally well.

Instructors, too, may want to integrate some literature with more of a disciplinary focus to expose students to the particular terminology and research methods that their discipline uses to analyze the political economy of global environmental change. One of us, for example, teaches in a department of political science and supplements this book with readings that reflect the language and debates in the fields of international relations and global environmental politics. The other teaches in both environmental and international development studies and supplements the book with readings that reflect the learning of the students in these programs. It is, we believe, worthwhile to encourage students to think beyond disciplinary boundaries. Yet often it is just as valuable to embed some learning within one or two disciplines, because this can allow for a more erudite analysis of the core questions in a particular discipline.

We trust all who choose to continue—regardless of the reason for beginning—will read with the curiosity of a true student, so each of the worldviews can spring equally to life in the analysis in the rest of the book.

Acknowledgments

We would like to thank all who have inspired us in one way or another with this book. For invaluable research assistance, special thanks go to Sam Grey for her tireless work and far-reaching talents. Sharon Goad, Joshua Gordon, Jeca Glor-Bell, and Sanushka Mudaliar also helped immeasurably with the research. We are grateful to the Australian Research Council and the Social Sciences and Humanities Research Council of Canada for financial support for this research. For perceptive comments and valuable guidance we would like to thank Herman Daly, Catherine Dauvergne, Torben Drewes, Derek Hall, Eric Helleiner, and three anonymous reviewers for The MIT Press. We are grateful to Clay Morgan for shepherding this book through the publication process. And we would like to thank our friends and families for their moral support and patience throughout this project. We accept full responsibility for any errors or omissions that are bound to arise in a book of this scope, and would appreciate any feedback from our readers.

Acronyms

AIDS	Acquired immune deficiency syndrome
AoA	Agreement on Agriculture
APEC	Asia Pacific Economic Cooperation
BAN	Basel Action Network
BASD	The Business Action on Sustainable Development
BECC	Border Environmental Cooperation Committee
BIS	Bank for International Settlements
CBD	Convention on Biological Diversity
CDIC	Canadian Development Investment Corporation
CEC	Commission on Environmental Cooperation
CEO	Chief executive officer
CFCs	Chlorofluorocarbons
CIDA	Canadian International Development Agency
CIS	Commonwealth of Independent States
CITES	Convention on International Trade in Endangered Species of Wild Flora and Fauna
CO_2	Carbon dioxide
COP	Conference of the Parties
CPRs	Common property regimes
CSD	UN Commission on Sustainable Development
CSR	Corporate social responsibility
CTE	Committee on Trade and Environment
DAC	OECD Development Assistance Committee
DCSD	Danish Committee on Scientific Dishonesty

DDT	Dichlorodiphenyltrichloroethane
DPRK	Democratic People's Republic of Korea
ECAs	Export credit agencies
ECGD	Export Credit Guarantee Department (UK)
ECOSOC	UN Economic and Social Council
EFIC	Australian Export Finance and Insurance Corporation
EIA	Energy Information Administration
EKC	Environmental Kuznets curve
EPA	Environmental Protection Agency (U.S.)
ESCAP	UN Economic and Social Commission for Asia and the Pacific
EU	European Union
FAO	Food and Agriculture Organization
FCCC	UN Framework Convention on Climate Change
FDA	Food and Drug Administration (U.S.)
FDI	Foreign direct investment
FOE	Friends of the Earth
FOEI	Friends of the Earth International
FSC	Forest Stewardship Council
G-77	Group of 77
GAST	General Agreement on Sustainable Trade
GATS	General Agreement on Trade in Services
GATT	General Agreement on Tariffs and Trade
GDP	Gross domestic product
GEF	Global Environment Facility
GEMI	Global Environmental Management Initiative
GEO	Global Environment Outlook
GMOs	Genetically modified organisms
GNI	Gross national income
GNP	Gross national product
GPI	Genuine progress indicator
HDI	Human development index

HIPC	Heavily indebted poor countries
HIV	Human immunodeficiency virus
IBRD	International Bank for Reconstruction and Development
ICC	International Chamber of Commerce
ICSID	International Center for the Settlement of Investment Disputes
IDA	International Development Association
IFC	International Finance Corporation
IFG	International Forum on Globalization
IIC	International Insolvency Court
IISD	International Institute for Sustainable Development
IMF	International Monetary Fund
IPCC	Intergovernmental Panel on Climate Change
ISEW	Index of sustainable economic welfare
ISO	International Organization for Standardization
ITU	International Telecommunications Union
IUCN	International Union for Conservation of Nature and Natural Resources
JBIC	Japan Bank for International Cooperation
LETS	Local exchange trading systems
LPI	Living planet index
MAI	Multilateral Agreement on Investment
MARPOL	Convention for the Prevention of Pollution by Ships
MEA	Multilateral Environmental Agreement
MFN	Most favored nation
MIGA	Multilateral Investment Guarantee Agency
MMPA	Marine Mammal Protection Act (U.S.)
NAAEC	North American Agreement on Environmental Cooperation
NAFTA	North American Free Trade Agreement
NASA	National Aeronautics and Space Administration
NGO	Nongovernmental organization
NIEO	New International Economic Order

ODA	Official Development Assistance
OECD	Organization for Economic Cooperation and Development
OPEC	Organization of Petroleum Exporting Countries
OPIC	Overseas Private Investment Corporation (U.S.)
PCBs	Polychlorinated biphenyls
POPs	Persistent organic pollutants
PPMs	Production and processing methods
PPP	Polluter pays principle
PVC	Polyvinyl chloride
SAL	Structural adjustment loan
SAP	Structural adjustment program
SPS	Agreement on the Application of Sanitary and Phytosanitary Measures
SUV	Sports utility vehicle
TBT	Agreement on Technical Barriers to Trade
TNC	Transnational corporation
TRIPS	Trade Related Intellectual Property Rights Agreement
TWN	Third World Network
UK	United Kingdom
UN	United Nations
UNCED	UN Conference on Environment and Development
UNCTAD	UN Commission on Trade and Development
UNCTC	UN Center on Transnational Corporations
UNDP	UN Development Programme
UNEP	UN Environment Programme
UNFPA	UN Fund for Population Activities
U.S.	United States
USAID	U.S. Agency for International Development
WBCSD	World Business Council for Sustainable Development
WCED	World Commission on Environment and Development
WDI	World Development Indicators
WEO	World Environment Organization

WHO	World Health Organization
WICE	World Industry Council on the Environment
WLO	World Localization Organization
WRI	World Resources Institute
WWF	WWF Network (formerly World Wildlife Fund/World Wide Fund for Nature)
WWW	World Wide Web

Paths to a Green World

1

Peril or Prosperity? Mapping Worldviews of Global Environmental Change

The sun could well engulf the earth in about 7 or 8 billion years. "So what," you might shrug. "The extinction of earth, beyond the horizon of human time, ridiculous, not worth imagining." Yet some environmentalists believe that waves of smaller disasters—like global warming, deforestation, and biodiversity loss—are already destroying the planet. Without doubt, too, many of the world's poorest people have already collided with their sun, dying from disease, starvation, war, and abuse. The beginning of the end, these environmentalists lament, is already upon us. We, as a species, are now beyond the earth's carrying capacity, a trend accelerating in the era of globalization. Unless we act immediately with resolve and sacrifice, in a mere hundred years or so, humanity itself will engulf the earth. The future is one of peril.

Many environmentalists rebel against such catastrophic visions. Yes, there are undeniable ecological problems—like the depletion of the ozone layer, the pollution of rivers and lakes, and the collapse of some fish stocks—but some ecological disturbance is inevitable, and much is correctable through goodwill and cooperation. There is no crisis or looming crisis: to think so is to misread the history of human progress. This history shows the value of positive thinking, of relying on human ingenuity to overcome obstacles and create ever-greater freedom and wealth with which we can ensure a better natural environment. Globalization is merely the latest, though perhaps the most potent, engine of human progress. The future is one of prosperity.

Who is correct? Do the pessimists need Prozac? Do the optimists need a stroll through a toxic waste dump in the developing world? Less flippantly, what is the middle ground between these two extremes? What are the causes and consequences of global environmental change? Are ecological problems really as severe as some claim? Does the cumulative impact of these problems constitute a crisis? How is the global community handling them? Why are the efforts to resolve some problems more successful than others? Why are environmental problems worse in some parts of the world? And what is the relationship to global political and economic activity? These are tough questions, and we do not pretend to know the answers with absolute certainty. A quick survey of the typical answers to these questions reveals an almost endless stream of contradictory explanations and evidence. Each answer can seem remarkably logical and persuasive. The result for the thoughtful and "objective" observer is often dismay or confusion.

Given this, how does one even begin to understand global environmental change? It helps, we believe, to begin with the big picture, rather than delving immediately into in-depth studies of particular environmental issues. Understanding this big picture is, in our view, necessary *before* we can fully understand the various interpretations of the *specific* causes and consequences of environmental problems. In the quest for knowledge and a role in a world overloaded with information and experts, far too often this larger picture is ignored—or at least poorly understood. For problems as intricate as global environmental ones, this can lead to muddled analysis and poorly formulated recommendations. Without this broad perspective, for example, "solving" one problem can ignore other related problems, or create even greater problems elsewhere.

How polities and societies allocate financial, human, and natural resources directly influences how we manage local, national, and ultimately global environments. The issues that shape the relationship between the global political economy and the environment are, of course, often technical and scientific. But they are frequently also socioeconomic and political. Our hope is that by sketching the arguments and assumptions about socioeconomic and political causes with the broadest possible strokes, we will assist readers in a lifelong journey of understanding the causes and consequences of global environmental change, as well as the controversies that surround it. This is a small yet essential step to eventually solving, or at least slowing, some of these problems.[1] To introduce these topics, we map out a new typology of worldviews on the political economy of global environmental change.[2]

Four Environmental Worldviews

We present four main worldviews on global environmental change and its relationship to the global political economy: those of *market liberals*, *institutionalists*, *bioenvironmentalists*, and *social greens*. These labels are intentionally transdisciplinary. Many books on the global environment confine the analysis to one disciplinary box—by limiting it, say, to political science theories or to economic models. This leaves far too many questions badly answered and far too many questions unasked. But we have had to make some choices. It is, of course, impossible to cover all disciplinary perspectives in one book. In our case we have chosen to rely mostly on the tools of political science, economics, development studies, environmental studies, political geography, and sociology. This focus, we believe, is narrow enough to do justice to the literature in these disciplines while still broad enough to provide new insights into the sources of environmental change and the possible options—both theoretical and practical—for managing it.

These are "ideal" categories, exaggerated to help differentiate between them. They are designed as tools to help simplify a seemingly unmanageable avalanche of conflicting information and analysis. Within each category, we have tried to clump thinkers—not just academics, but equally policymakers and activists—with broadly common assumptions and conclusions. This we hope provides a sense of the debates in the "real" world—that is, within bureaucracies, cabinet meetings, international negotiations, and corporate boardrooms, as well as in classrooms. Our approach, in a sense, tries to capture the societal debates about environment and political economy rather than just the academic debates over the theories of the political economy of the environment.

Naturally, given the breadth of our labels, many disagreements exist among those in each category. We have tried to show the range of views subsumed under each of the four major worldviews, although at the end of this book you may still find that your own beliefs and arguments do not fit neatly into any of these categories. Or you may feel that you hold a mix of views—even ones that at first seem at opposite poles, such as market liberal and social green. This does not mean that our categories are erroneous. Or that you are inconsistent or hypocritical. Or that you should force your views into one category. Instead, it just shows the complexity and diversity of individual views on the issues.

Our typology, moreover, does not cover all possible views, although, while conscious to avoid creating dozens of labels, we do try to give a reasonable range. We only include thinkers who are *environmentalists*—that is, those who write and speak and work to maintain or improve the environment around us. This includes those highly critical of so-called environmental activists or radical greens. An economist at the World Bank is, in our view, just as much an environmentalist as a volunteer at Greenpeace, as long as the economist believes she or he is working for a better environment (however that is defined). Also, we focus principally on economic and political arguments, and tend to give less attention to philosophical and moral ones. Within the political and economic literature, we stress arguments and theories that try to *explain* global environmental change—that is, the literature that looks at an environmental problem and asks: Why is that happening? What is causing it? And what can be done?

With those introductory remarks, we now turn to our typology.

Market Liberals

The analysis of market liberals is grounded in neoclassical economics and scientific research. Market liberals believe that economic growth and high per capita incomes are essential for human welfare and the maintenance of sustainable development. Sustainable development is generally defined by these thinkers along the lines of the 1987 World Commission on Environment and Development (WCED): "development that meets the needs of the present without compromising the ability of future generations to meet their own needs."[3] In terms of improving global environmental conditions, market liberals argue that economic growth (production and consumption) creates higher incomes, which in turn generates the funds and political will to improve environmental conditions. Rapid growth may exacerbate inequalities, as some of the rich become super rich, but in the long run all will be better off. In other words, all boats will rise. Market-liberal analysis along these lines is commonly found, for example, in publications of the World Bank, the World Trade Organization (WTO), and the World Business Council for Sustainable Development (WBCSD), as well as in the media in publications such as *The Economist.*

Market liberals see globalization as a positive force, because it promotes economic growth as well as global integration. They concede that as states pursue economic growth, environmental conditions—such as air and water quality—may deteriorate as governments and citizens give firms more scope to pursue short-term profits, thus stimulating further economic growth. But once a society becomes wealthy, citizens (and in turn governments and business) will raise environmental standards and expectations. *The Economist* magazine explains the global pattern: "Where most of the economic growth has occurred—the rich countries—the environment has become cleaner and healthier. It is in the poor countries, where growth has been generally meagre, that air and water pollution is an increasing hazard to health."[4] The key, market liberals argue, is good policy to ensure that economic growth improves the environment in all countries.

The main drivers of environmental degradation, according to market liberals, are a lack of economic growth, poverty, distortions and failures of the market, and bad policies. The poor are not viewed as unconcerned or ignorant. Rather, to survive—to eat, to build homes, to earn a living—they must exploit the natural resources around them. They are, according to the World Bank, both "victims and agents of environmental damage."[5] It is unrealistic, perhaps even unjust, to ask the poor to consider the implications of their survival for future generations. The only way out of this vicious cycle is to alleviate poverty, for which growth is essential. Restrictive trade and investment policies and a lack of secure property rights all hamper the ability of the market to foster growth and reduce poverty. Market failures—instances where the free market results in an environmentally suboptimal outcome—are viewed as possible causes of some environmental problems, although these are seen as relatively rare in practice. More often, market liberals argue, poor government policies—especially those that distort the market, such as subsidies—are the problem.

Market liberals frequently draw on more moderate estimates of environmental damage and more optimistic scenarios for the future. A few have become famous for declaring that the global environment is nowhere near a state of crisis—such as economist Julian Simon,[6] popular columnist Gregg Easterbrook,[7] and political scientist Bjørn Lomborg.[8] But most recognize that many environmental problems are indeed serious, although all reject the image of the world spinning toward a catastrophic ecological crash. Instead market liberals tend to stress our scientific achievements, our progress, and our ability to reverse and repair environmental problems with ingenuity, technology, cooperation, and adaptation. For these thinkers, population growth and resource scarcity are not major concerns when it comes to environmental quality. A glance at the historical trend of better environmental conditions for all confirms this (especially statistics from the developed world). So do the global data on human well-being, such as medical advances, longer life expectancy, and greater food production. Furthermore, most environmental problems, if not currently responding to efforts to manage them more effectively, at least have the potential to improve in the longer term.

Thinkers from the market-liberal tradition place great faith in the ability of modern science and technology to help societies slip out of any environmental binds that may occur (if, for example, there are unavoidable market failures). Human ingenuity is seen to have no limits. If resources become scarce, or if pollution becomes a problem, humans will discover substitutes and develop new, more environmentally friendly technologies. Market liberals see advances in agricultural biotechnology, for example, as a key answer to providing more food for a growing world population. Their belief in science leaves most market liberals wary of precautionary policies that restrict the use of new technology, unless there is clear scientific evidence to demonstrate it is harmful.

Market liberals believe open and globally integrated markets promote growth, which in turn helps societies find ways to improve or repair environmental conditions. To achieve these goals market liberals call for policy reforms to liberalize trade and investment, foster specialization, and reduce government subsidies that distort markets and waste resources. Governments, too, need to strengthen some institutions, such as institutions to secure property rights or institutions to educate and train the poor to protect the environment. Governments are encouraged to use market-based tools—for example, environmental taxes or tradable pollution permits—to correct situations of genuine market failure. Innovative environmental markets—like a global scheme to trade carbon emissions or niche markets for environmental products such as timber from sustainable sources—and voluntary corporate measures to promote environmental stewardship are also reasonable ways to improve environmental management. But in most cases it is best to let the market allocate resources efficiently. Market liberals, such as the economist Jagdish Bhagwati[9] and the business executive Stephan Schmidheiny,[10] strongly argue that it makes economic sense for firms to improve their environmental performance, and for this reason it makes sense to let the market guide them.

Institutionalists

The ideas of institutionalists are grounded in the fields of political science and international relations. They share many of the broad assumptions and arguments of market liberals—especially the belief in the value of economic growth, globalization, trade, foreign investment, technology, and the notion of sustainable development. Indeed, moderate institutionalists sit close to moderate market liberals. It is a matter of emphasis. Market liberals stress more the benefits and dynamic solutions of free markets and technology; institutionalists emphasize the need for stronger global institutions and norms as well as sufficient state and local capacity to constrain and direct the global political economy. Institutions provide a crucial route to transfer technology and funds to the poorest parts of the planet.[11] Institutionalists also worry far more than market liberals about environmental scarcity, population growth, and the growing inequalities between and within states. But they do not see these problems as beyond hope. To address them, they stress the need for strong institutions and norms to protect the common good. Institutionalist analysis is found in publications by organizations such as the United Nations Environment Programme (UNEP) and by many academics who focus their analysis on "regimes" (international environmental agreements and norms, defined more precisely in chapter 3) in the fields of political science and law.

Institutionalists see a lack of global cooperation as a key source of environmental degradation. Ineffective cooperation partly arises because of the nature of the sovereign state system, which gives a state supreme authority within its boundaries. In such a system states tend to act in their own interest, generally leaving aside the interest of the global commons. Yet like market liberals, institutionalists *do not* reject the way we have organized political and economic life on the planet. Instead they believe we can overcome the problem of sovereignty as the organizing principle of the international system by building and strengthening global and local institutions that promote state adherence to collective goals and norms. This can be most effectively carried out through global-level environmental agreements and organizations.

The process of globalization makes global cooperation increasingly essential (and increasingly inevitable). But institutionalists stress that unfettered globalization can add to the pressures on the global environment. The task for those worried about the state of the global environment, then, is to guide and channel globalization, so it enhances environmental cooperation and better environmental management. This point has been stressed most forcefully by key policy figures such as former Norwegian Prime Minister Gro Harlem Brundtland in her role in the 1980s as head of the World Commission on Environment and Development and Canadian diplomat Maurice Strong as organizer of global environmental conferences. The aim of this approach is to ensure that global economic policies work to both improve the environment and raise living standards.[12] Controls at all levels of governance, from the local to the national to the global, can help to direct globalization, enhancing the benefits and limiting the drawbacks.[13]

For the global environment, institutionalists believe that institutions need to internalize the principles of sustainable development, including into the decision-making processes of state bureaucracies, corporations, and international organizations. Only then will we be able to manage economies and environments effectively—especially for common resources. For many institutionalist academics, like political scientist Oran Young, the most effective and practical means is to negotiate and strengthen international environmental regimes.[14] Many within the policy world, such as in the United Nations Environment Programme, add the need to enhance state and local capacity in developing countries.[15] Thus, many institutionalists call for "environmental aid" for the developing world.[16] It should be stressed, however, that institutionalists do not necessarily support all institutions uncritically. Some point to badly constructed institutions as a source of problems. Many point, too, to the difficulty of trying to measure the implementation and effectiveness of an international agreement or institution.[17] But a defining characteristic of institutionalists is the assumption that institutions matter—that they are valuable—and that what we need to do is reform, not overthrow, them.

Institutionalists also argue that strong global institutions and cooperative norms can help enhance the capacity of *all* states to manage environmental resources. What is needed, from this perspective, is to effectively embed environmental norms in international cooperative agreements and organizations as well as state policies. Along these lines, many institutionalists support the precautionary approach, where states agree to collective action in the face of some scientific uncertainty. Institutionalists also advocate the transfer of knowledge, finances, and technology to developing countries. Organizations like the World Bank, the United Nations Environment Programme, and the Global Environment Facility (GEF) already play a role here. And many institutionalists point to the creation of, and changes within, these organizations as evidence of progress.

Bioenvironmentalists

Inspired by the laws of physical science, bioenvironmentalists stress the biological limits of the earth to support life. The planet is fragile, an ecosystem like any other. Some even see the earth as behaving like a living being, a self-regulating, complex, and holistic superorganism—the so-called Gaia hypothesis, as articulated by environmental scientist James Lovelock.[18] The earth can support life, but only to a certain limit, often referred to as the earth's "carrying capacity." Many bioenvironmentalists see humans as anthropocentric and selfish (or at least self-interested) animals. Some, like the academic William Rees, even see humans as having "a genetic predisposition for unsustainability."[19] All bioenvironmentalists agree that humans as a species now consume far too much of the earth's resources, such that we are near, or indeed have already overstepped, the earth's carrying capacity. Such behavior, without drastic changes, will push the planet toward a fate not much different from the ecological calamity of Easter Island of 300 years ago—where a once-thriving people became over a few centuries "about 2000 wretched individuals . . . eking out a sparse existence from a denuded landscape and cannibalistic raids on each other's camps."[20] These scholars stress the environmental disasters around us, often citing shocking figures on such problems as overfishing, deforestation, species loss, and unstable weather patterns. Publications of the Worldwatch Institute and the WWF are illustrative of this perspective.

For most bioenvironmentalists population growth is a key source of stress on the earth's limits. The ideas of Thomas Malthus (1766–1834), who in "An Essay on the Principle of Population"[21] predicted that the human population would soon outstrip food supply, were revived in the late 1960s by writers such as biologist Paul Ehrlich.[22] Sometimes known as neo-Malthusians, these writers argue that global environmental problems ultimately stem from too many people on a planet with finite resources. The principle of sovereignty, which divides the world into artificial territories, aggravates the effects of too many humans, because it violates the principles of ecology and creates what academic Garrett Hardin famously called a "tragedy of the commons." For him, too many people without overarching rules on how to use the commons creates a situation where individuals, rationally seeking to maximize their own gain at the expense of others, overuse and ultimately destroy the commons.[23] This point, stressed by many bioenvironmentalists, is also made by many institutionalists, as discussed earlier.

Many bioenvironmentalists stress, too, that the neoclassical economic assumption of infinite economic growth is a key source of today's global environmental crisis. For these thinkers, a relentless drive to produce ever more in the name of economic growth is exhausting our resources and polluting the planet. Many argue that the drive to pursue ever more economic growth is what has taken the earth beyond its carrying capacity. For bioenvironmentalists, human consumption patterns are as great a problem as population growth, and the two are seen as inextricably linked.[24] They argue that together rising populations and consumption are drawing down the earth's limited resources, and that we must respect the biophysical limits to growth: both for people and economies.

Not all bioenvironmentalists engage directly in discussions on economic globalization, but those that do tend to see globalization as a negative force for the environment. They agree with market liberals that globalization enhances economic growth. But instead of seeing this as positive for the environment, they see it as contributing to further environmental degradation. For them, more growth only means more consumption of natural resources and more stress on waste sinks. Globalization is blamed, too, for spreading Western patterns of consumption into the developing world. With much larger populations and often more fragile ecosystems (especially in the tropics), this spread of consumerism is accelerating the collapse of the global ecosystem.[25] Globalization is also seen to encourage environmentally harmful production processes in poor countries that have lower environmental standards.[26] For these reasons, these bioenvironmentalists argue that we must curtail economic globalization to save the planet.

Solutions proposed by bioenvironmentalists flow logically from their analysis of the causes of environmental damage: we need to curb economic and population growth. Those who focus on the limits to economic growth have been a core group in the field of ecological economics, pioneered by thinkers such as the economist Herman Daly[27] and published in journals such as *Ecological Economics*. This group combines ideas from the physical sciences and economics to develop proposals to revamp economic models to include the notion of physical limits, which involves changing our measures of "progress" and the methods we use to promote it. Only then, these thinkers argue, can we reduce the impact of humans on the planet and prod the world toward a more sustainable global economy. Those bioenvironmentalists who focus more on overpopulation call for measures to lower population growth, like expanding family planning programs in the Third World, and for curbs on immigration to rich countries where consumption problems are the worst. At the more extreme end, some see a world government with coercive powers as the best way to control the human lust to fill all ecological space, destroying it, often inadvertently, in the process.[28]

Social Greens

Social greens, drawing primarily on radical social and economic theories, see social and environmental problems as inseparable. Inequality and domination, exacerbated by economic globalization, are seen as leading to unequal access to resources as well as unequal exposure to environmental harms. While these views have long been important in debates over environment and development, and are themselves a mix of a variety of radical views, scholars in international political economy have only recently recognized them as a distinct perspective.[29]

Many social greens from a more activist stance focus on the destructive effects of the global spread of large-scale industrial life.[30] Accelerated by the process of globalization, large-scale industrialism is seen to encourage inequality characterized by overconsumption by the wealthy while at the same time contributing to poverty and environmental degradation. While agreeing broadly with this analysis, other, more academic social greens draw on Marxist thought, pointing specifically to capitalism as a primary driver of social and environmental injustice in a globalized world. They argue that capitalism, and its global spread via neocolonial relations between rich and poor countries, not only leads to an unequal distribution of global income, power, and environmental problems, but is a threat to human survival.[31] Also inspired by Marxist thought, some social greens take a neo-Gramscian, or historical materialist perspective, focusing on the way those in power frame and influence ecological problems, primarily hegemonic blocs consisting of large corporations and industrial country governments.[32] Other social greens like Vandana Shiva draw heavily from feminist theory to argue that patriarchal relationships in the global economy are intricately tied to ecological destruction.[33] The key concern of all of these strands of social green thought, then, is inequality and the environmental consequences related to it. Social green analysis can be found in magazines such as *The Ecologist* and in reports of groups such as the International Forum on Globalization (IFG) and the Third World Network (TWN).

Social greens sympathize with bioenvironmentalist arguments that physical limits to economic growth exist. Overconsumption, particularly in rich industrialized countries, is seen by social greens to put a great strain on the global environment.[34] Many, perhaps most prominently Wolfgang Sachs[35] and Edward Goldsmith,[36] see this problem as accelerating in an era of economic globalization. The arguments of social greens on growth and consumption, and on the role of the global economy in accelerating both, are close to bioenvironmentalist arguments. But few social greens accept bioenvironmentalist arguments regarding population growth, instead maintaining that overconsumption, particularly among the rich in the First World, is a far greater problem.[37] Unlike bioenvironmentalists, most social greens see population-control policies as a threat to the self-determination of women and the poor.[38]

Whether it is viewed as spreading industrialism or capitalism (or both), social greens uniformly oppose economic globalization, arguing that it is a key factor behind much of what is wrong with the global system.[39] In addition to feeding environmentally destructive growth and consumption, globalization is seen to breed injustice in a number of ways. It exacerbates the inequality within and between countries. It reinforces the domination of the global rich and the marginalization of women, indigenous peoples, and the poor. It assists corporate exploitation of the developing world (especially labor and natural resources). It weakens local community autonomy and imposes new forms of domination that are Western and patriarchal (local customs, norms, and knowledge are lost, replaced by new forms unsuited to these new locations). Globalization is also seen to destroy local livelihoods, leaving large numbers of people disconnected from the environment in both rich and poor countries. This globalization is viewed by many social greens as a continuation of earlier waves of domination and control. In the words of the prominent antiglobalization activist Vandana Shiva, "The 'global' of today reflects a modern version of the global reach of the handful of British merchant adventurers who, as the East India Company, later, the British Empire raided and looted large areas of the world."[40]

From this analysis, it is not surprising that social greens reject the current global economy. Reactive crisis management in a globalized world, social greens believe, will not suffice to save the planet: tinkering will just momentarily stall the crash. In many instances the environmental solutions of market liberals and institutionalists, because they assume globalization brings environmental benefits, are part of the problem. For social greens major reforms are necessary, well beyond, for example, just strengthening institutions or internalizing environmental and social costs into the price of traded goods. Thus social greens, as the work of the International Forum on Globalization exemplifies, call for a dismantling of current global economic structures and institutions.[41] To replace this, many social greens advocate a return to local community autonomy to rejuvenate social relations and restore the natural environment. Localization activist Colin Hines has mapped out a model for

Table 1.1
Environmental perspectives

	Market liberals	Institutionalists	Bioenvironmentalists	Social greens
Focus	Economies	Institutions	Ecosystems	Justice
A global environmental crisis?	No. Some inevitable problems, but overall modern science, technology, ingenuity, and money are improving the global environment.	Not yet. Potential for crisis unless we act now to enhance state capacity and improve the effectiveness of regimes and global institutions.	Yes. Near or beyond earth's carrying capacity. Ecological crisis threatens human survival.	Yes. Social injustice at both local and global levels feeds environmental crisis.
Causes of problems	Poverty and weak economic growth. Market failures and poor government policy (i.e., market distortions such as subsidies as well as unclear property rights) are also partly to blame.	Weak institutions and inadequate global cooperation to correct environmental failures, underdevelopment, and perverse effects of state sovereignty.	Human instinct to overfill ecological space, as seen by overpopulation, excessive economic growth, and overconsumption.	Large-scale industrial life (some say global capitalism), which feeds exploitation (of labor, women, indigenous peoples, the poor, and the environment), and grossly unequal patterns of consumption.

Impact of globalization	Fostering economic growth, a source of progress that will improve the environment in the long run.	Enhancing opportunities for cooperation. Guided globalization enhances human welfare.	Driving unsustainable growth, trade, investment, and debt. Accelerating depletion of natural resources and filling of waste sinks.	Accelerating exploitation, inequalities, and ecological injustice while concurrently eroding local-community autonomy.
The way forward	Promote growth, alleviate poverty, and enhance efficiency, best pursued with globalization. Correct market and policy failures, and use market-based incentives to encourage clean technologies. Promote voluntary corporate greening.	Harness globalization and promote strong global institutions, norms, and regimes that manage the global environment and distribute technology and funds more effectively to developing countries. Build state capacity. Employ precautionary principle.	Create a new global economy within limits to growth. Limit population growth and reduce consumption. Internalize the value of nonhuman life into institutions and policies. Agree to collective coercion (e.g., some advocate world government) to control greed, exploitation, and reproduction.	Reject industrialism (and/or capitalism) and reverse economic globalization. Restore local community autonomy and empower those whose voices have been marginalized. Promote ecological justice and local and indigenous knowledge systems.

how this could occur. It entails a retreat from the large-scale industrial and capitalist life and a move toward local, self-reliant, small-scale economies.[42] These thinkers stress the need to, in the words of some, "think globally, act locally." In other words, understand the global context, while at the same time acting in ways suitable to the local context. These thinkers advocate bioregional and small-scale community development—because they firmly believe that a stronger sense of community will fulfill basic needs and enhance people's quality of life. Such development would help to reduce inequities and levels of consumption that are out of balance with the world's natural limits.

As part of their strategy for promoting community autonomy and localization, social greens also stress the need to empower those voices that the process of economic globalization is marginalizing. They embrace indigenous knowledge systems, for example, arguing these are equally if not more valid than the Western scientific method. The process of economic "development," these critics argue, foists the latter onto the developing world, thus threatening ecologically sound local systems.[43] Many social greens regard local cultural diversity as essential for the maintenance of biological diversity. The erosion of one is seen to lead to the erosion of the other. In advocating local and indigenous empowerment and input, social greens emphasize that effective solutions to environmental problems will continue to remain elusive unless the voices of women, indigenous peoples, and the poor are integrated into the global dialogue on environment and development, as well as into locally specific contexts.

Conclusion

Table 1.1 summarizes the main assumptions and arguments of market liberals, institutionalists, bioenvironmentalists, and social greens. We have tried hard to present these views fairly and accurately. Yet we also stress again that these are "ideal" categories, and within each there are a range of views and more subtle debates. Some authors you will read will fit neatly into one of these categories, while others are more difficult to classify. This just demonstrates the range of possible views. Moreover, there are alliances between various views on different issues, which makes the terrain difficult to map at times. For example, market liberals and institutionalists agree with one another that economic growth and globalization have positive implications for the environment, while social greens and bioenvironmentalists hold the opposite view. And institutionalists and bioenvironmentalists agree that population growth poses a problem for the world's resources, while market liberals and social greens put far less emphasis on this factor.

We stress that we do not want to leave the impression that any one of these is the "correct" view. Each, we believe, contains insights into the sources of today's environmental problems, as well as into potential solutions. Each view has its own logic, which fits with its assumptions. Understanding these views help to explain, too, the often markedly different interpretations of the condition of the global environment. One article, for example, may well declare global warming the most serious threat confronting today's governments. The next article may declare it a hoax, a ploy to raise funds or perhaps to scare world leaders into action. This, we believe, does not mean that there are no facts—or causality—or analysis—or statistics. It also does not mean that some authors lie and deceive. Rather it merely shows how different interpretations and different values—that is, different worldviews—can shape which information an analyst chooses to *emphasize*.

As you proceed through the rest of this book, we urge you to keep an open mind regarding the debates and evidence about the consequences of the global political economy for global environmental change. This is certainly not easy. These are emotional issues. And the evidence and arguments are often contradictory, almost as if analysts live in different worlds. Our hope, if you do keep an open mind until the end of the book, is not to confuse you, but to leave you with a better understanding of your own assumptions and arguments. Moreover, if you then decide to reject the arguments of others, you will do so with a genuine understanding of the complexity and historical sources of those views. Only then can the debates truly move forward.

2

The Ecological Consequences of Globalization

We are now in an era of globalization. As a process that touches on many aspects of our economies, societies, and cultures, it is important to investigate how it interfaces with global environmental change. This chapter begins with a general overview of globalization—what it is and its implications. It then sketches broadly contrasting environmental pictures of today's globalizing world: first, a world of progress, of better lives that result from globalization, the world as seen by market liberals and to a lesser extent institutionalists; and second, a world of failure, of crisis and looming ecological and social catastrophe unless immediate action is taken to reverse globalization, the world as seen by bioenvironmentalists, and many social greens. The first picture stresses the social and environmental benefits of macroeconomic growth, capitalism, new technologies, and international cooperation. The second picture stresses the biological strains of overpopulation and unequal consumption on a planet with too much industrial and agricultural production. Within each broad picture, however, the different perspectives emphasize different points. The chapter concludes with a sketch of how the different lenses of the market liberals, institutionalists, bioenvironmentalists, and social greens provide different insights into the process of globalization. Chapter 3 will then more specifically address globalization and the development of global environmental discourse, as well as the evolution of environmental actors, institutions, and norms at the international level.

What Is Globalization?

Some scholars question whether globalization is actually occurring. Others question whether the word has any real meaning. Still others see the definitions that swirl around the term as analytically imprecise. Without doubt the concept of globalization is vaguely and imprecisely defined at times, and some authors seem to use the concept inconsistently. In spite of these difficulties, we believe it is still a valuable concept and helps to illuminate the relationship between global environmental change and the global political economy. In fact, we rather boldly assert that globalization is a critical force shaping global affairs.[1] Rather than getting bogged down in the debate over the various possible definitions of globalization, we will just outline our understanding of the term.

Globalization is a multidimensional process, broadly restructuring and integrating the world's economies, institutions, and civil societies. It is a dynamic, ongoing, and accelerating process that is increasing the links among actors, as well as the structures within which they operate, both within states and across borders. Trade, production, and finance are now more globally integrated than ever before, as are global organizations and social movements. These extensive linkages are making global interactions more intense and complex. Globalization is also changing the nature of space, collapsing some space—for example, the effective distance between individuals in Africa and Asia—while creating sharper spatial barriers elsewhere—for instance, between neighbors on a suburban street, customers in a megasupermarket, and commuters on a subway. It is also affecting time, speeding up the nature of these interactions as communication hardware advances. In simple terms globalization means that the events and actions in one part of the world are affecting people in distant lands much more quickly, and with greater frequency and intensity.

Globalization is partly an extension of processes that began long ago, including modernization and colonization. But it is more than just a new term for old phenomena. It is also more than just internationalization, which focuses on the importance of greater economic and political interconnectedness among states, and which can be seen as a subprocess of globalization. It suggests instead "that human lives are increasingly played out in *the world as a single place.*"[2] It points to the growing significance of a global community with global concerns, and it stresses the decreasing importance of geographic distance and increasing importance of transnational actors and forces. It further suggests that we are moving toward an effectively borderless world (especially for ideas and money), and that this process is accelerating (see box 2.1 for selected definitions of globalization).

If this is globalization, what is driving it? Separating the drivers from the consequences of globalization is difficult, because the consequences are themselves constantly reshaping the drivers. New and faster communication technologies—such as telephones, faxes, and e-mail—partly drive it. These technologies create a faster and more efficient transmission belt for people, money, and ideas, as well as for the knowledge to

Box 2.1
Definitions of globalization

> *New York Times* columnist Thomas Friedman writes that "globalization is the integration of everything with everything else. . . . [It] is the integration of markets, finance, and technology in a way that shrinks the world from a size medium to a size small. Globalization enables each of us, wherever we live, to reach around the world farther, faster, deeper, and cheaper than ever before and at the same time allows the world to reach into each of us farther, faster, deeper, and cheaper than ever before."[1]
>
> Political scientist David Held and his colleagues define globalization initially as "the widening, deepening and speeding up of worldwide interconnectedness in all aspects of contemporary social life, from the cultural to the criminal, the financial to the spiritual." Later they add more precision, offering this definition: "a process (or set of processes) which embodies a transformation in the spatial organization of social relations and transactions—assessed in terms of their extensity, intensity, velocity and impact—generating transcontinental or interregional flows and networks of activity, interaction, and the exercise of power."[2]
>
> Economist John Helliwell notes: "For protestors in Seattle, Gothenberg, and Genoa, globalization represents a state of the world wherein international organizations implement the wishes of transnational corporations, ensuring that free trade rules will combine with global market pressures to eliminate the ability of local and national governments to implement policies. . . . Within corporations and business groups, globalization usually refers to a global market reach and to an imperative that firms must 'globalize or die.' . . . Among economists, globalization refers to a situation where the so-called 'law of one price' applies on a global basis. This assumes that goods and services will be freely and costlessly traded over space and borders."[3]
>
> 1. Friedman 2002.
> 2. Held et al. 1999, 2, 16.
> 3. Helliwell 2002, 15–16.

build other technologies—such as cars and industrial equipment. The first transborder telephone call was in 1891, between London and Paris. Since then the number of telephone mainlines has grown steadily, from 150 million in 1965, to 546 million in 1991, to 1,053 million in 2001.[3] Over the last hundred years the cost of international phone calls fell precipitously, from US$235 for a three-minute call from New York to London in 1930 to just 35 cents in 1998 (in constant 1990 U.S. dollars).[4] With more lines and cheaper rates the volume of international calls continues to rise steadily, from 12.7 billion minutes spent on the phone in 1982 to 127 billion in 2001.[5] The Internet is the latest revolutionary change in global communication. The World Wide Web had just 50 pages in 1993; by the end of the decade there were 50 million.[6] The number of Internet users went from 25 million in 1995 to 400 million by the end of 2000. The UN Development Programme (UNDP) predicts this will rise to 1 billion by 2005.[7] Perhaps most remarkable is the increase in the volume and speed of information flows on the Internet. By 2001 one Internet cable carried more information in a second than was sent over the whole Internet in a month in 1997.[8]

Faster and cheaper transportation facilitates communication as well. Here too there have been remarkable changes over the past half century. In 1950 there were 25 million air passengers; by 1996 there were 400 million; by 2000 1.4 billion. Part of the reason for this growth in air travel is that it is now much cheaper. In constant 1990 U.S. dollars, the cost fell from $0.68 per passenger mile in 1930 to $0.11 in 1990.[9]

Globalization is driven, too, by the increasing dominance of capitalism and Western ideologies, which stress the value of global economic integration. States themselves try to embed these values in global norms and institutions, in the hope that this will enhance their economic position in the global economy. A host of nongovernmental and intergovernmental organizations, as well as the global media and multinational corporate networks—involving a complex array of both reinforcing and contradictory pressures—collectively enhance global linkages. Global economic institutions and norms reflecting these values encourage further global integration of trade, production, and financial relationships.

Further, global environmental problems themselves, such as climate change, ozone depletion, and species extinction, also strengthen global consciousness. Because such problems do not respect borders, they are by definition "global-level" problems that require global solutions. Conferences such as the 1972 UN Conference on the Human Environment (Stockholm Conference) and the 1992 UN Conference on Environment and Development (Rio Conference), which focused attention on global environmental problems, raised the profile of these issues among the world's political elites (for details, see chapter 3).

When did globalization begin? We cannot answer this question precisely, because the process is an extension of historical, political, and socioeconomic processes that started long ago. It is reasonable, however, to point to the 1960s as the beginning of the rapid intensification of the process of contemporary globalization. Political scientist Jan Aart Scholte would agree here, stating that "globalization did not figure continually, comprehensively, intensely, and with rapidly increasing frequency in the lives of a large proportion of humanity until around the 1960s."[10]

Economic trends since the 1960s reveal the extent of globalization. Cross-border financial flows, foreign direct investment, and international trade have grown phenomenally, as governments have actively removed barriers to these types of transactions. About US$1.5 trillion in foreign currency is traded per day, an amount that is over ten times the value of underlying trade and investment flows, and significantly more than the average of US$10–20 billion in the 1970s.[11] The number of transnational parent firms has grown from 7,000 in 1970 to over 65,000 today, with over 850,000 affiliate firms. These firms make up one-tenth of world gross domestic product (GDP) and one-third of world exports.[12] The value of world merchandise exports was over US$6 trillion in 2000, over a hundred times the value of US$58 billion in 1948.[13]

Since the 1960s globalization has to some extent been promoting greater political, economic, cultural, and technological uniformity across the globe. Most governments now claim to be democratic[14] and many have introduced political structures similar to the ones in Western Europe and North America.[15] Most governments strive to integrate their domestic economic production into the global financial structure, introducing policies to foster foreign direct investment and trade, generally with an underlying goal of promoting higher rates of macroeconomic growth. A superficial glance at the cities around the world reveals striking similarities associated with this economic growth and with globalization: cars, roads, concrete, and steel. And consumer preferences are converging, with blue jeans, American sitcoms, and McDonald's just three of the many possible examples. The process of increasing uniformity, however, is far from inevitable. Fragmentation of whole countries and decentralization within particular countries are occurring alongside of and as part of globalization.

While some argue that globalization is an inevitable process, others argue that it is possible to resist it. Governments have played a role in fostering it. Similarly they can break global political ties or erect economic barriers. Activist and religious organizations, using some of the same technologies driving globalization, can disseminate alternative views in an effort to foster support for an alterative to globalization.[16] And groups within particular societies can reassert cultural identities and gain global support for their movement to do so.

Though globalization is in many ways a unifying force, its effects are uneven, both across and within countries, triggering great change in some places and virtually none in others. Some of these changes are positive, bettering people's everyday lives. But other changes are contributing to greater inequality in some areas (particularly in a relative sense) as well as to deeper levels of poverty in some areas (particularly in terms of access to sustainable livelihoods, nutrition, and community well-being). Distribution of global income is one indicator. The 1 billion who live in the rich industrialized countries receive 60 percent of global income; meanwhile, 3.5 billion in low-income countries receive less than 20 percent of global income. The wealthiest 20 percent of the world accounts for 86 percent of total private consumption expenditures. This does not just translate into far more luxury goods for the rich, like paper and cars. The richest 20 percent consume, too, 45 percent of all meat and fish. Also, technological research, including medical research, tends to concentrate on the needs of the wealthy. For example, from 1975 to 1996, of the more than 1,200 new drugs worldwide, only 13 treated tropical diseases.[17]

Access to the communication and transportation technologies of globalization is unequal, too. Of all international telephone traffic, 80 percent originates in the countries in the Organization for Economic Cooperation and Development (OECD, the thirty-member group representing the rich industrialized world). The OECD average international call usage is 36.6 minutes per person per year. In sub-Saharan Africa it is only 1 minute per person.[18] Countries like Cambodia, Uganda, and Bangladesh have just 3 telephone mainlines for every 1,000 people, while countries like Canada, Sweden, the United States, and Denmark have over 600 mainlines per 1,000 people.[19] Similar inequalities exist for personal computers and the Internet. Almost 80 percent of all Internet users live in OECD countries (14 percent of the global population). In 1999 Zimbabwe had 13 personal computers per 1,000 people, Thailand had 22.7 per 1,000 people, and the United States had 680 per 1,000 people.[20] In 2000, 54.3 percent of the U.S. population used the Internet, while only 0.4 percent did in South Asia and sub-Saharan Africa.[21]

There is also great inequality in the movement of people. The number of international tourists rose from 25 million in 1950 to 699 million in 2000. Of these, over half were from Europe; less than 3 percent were from Africa and less than 1 percent came from South Asia.[22] It is relatively easy, too, for the educated and the rich to immigrate. But it is now harder for poor people in poor countries who want to immigrate legally.[23] Such differences in access are contributing to a new form of inequality: the connected and unconnected. Using technologies to integrate into the global economic system provides numerous opportunities. Life can be tough, however, for those left unconnected—just look at the recent famines in North Korea and the violent conflict in the Congo over the past decade. Globalization is, in some ways, a "new Industrial Revolution" that, according to the UNEP, "could result in a dangerous polarization between people and countries benefiting from the system and those that are merely passive recipients of its effects."[24]

Globalization, by its very nature, is altering the underlying processes of global environmental change. Are the effects of these changes positive or negative for the overall environmental health of the planet? Before we examine the diverse range of answers to this question, it is useful to note that critics of globalization are often referring to economic globalization rather than to other forms of globalization. Many of the technologies that have facilitated economic globalization—the computer, for example—are seen as essential for the creation of global environmental norms and a global civil society, potential counterbalances to globalization that help citizens to "think globally and act locally." E-mail and the Internet are also cheap and effective tools of antiglobalization activists, enabling groups without overarching leadership to organize global protests from Seattle to Genoa to Bangkok on short notice. Many of the critics of economic globalization, although certainly not all, are actually supporters of what might be called social globalization.

Globalization and the Global Environment

The interpretation of the effects of globalization on the globe's natural environment evokes polar reactions. The first, reflective of how market liberals and institutionalists perceive the world, stresses our achievements, our progress over time, and our ability to promote economic well-being as well as reverse and repair environmental problems with ingenuity, technology, cooperation, and adaptation. The second, reflective of how bioenvironmentalists and social greens see the earth, emphasizes environmental disasters and human inequality, problems aggravated and in many cases driven by economic globalization.[25] We now sketch these two broad views of globalization. It should be stressed, though, that while there is a split at the broad level over whether the process of globalization is on the whole positive or negative, each worldview emphasizes different aspects of it.

Global Positive
Market liberals and institutionalists perceive globalization as an engine of wealth creation. Both argue that the globalization of trade, investment, and finance is pushing up global gross domestic product (GDP— for a detailed explanation of this measure of growth, see chapter 4) and global per capita incomes, which they see as essential to finance sustainable development. Over the period 1970–2000, one of heightened global economic integration, world GDP (in constant 1995 U.S. dollars) nearly tripled from US$13.4 trillion to US$34.1 trillion.[26] The annual per capita global growth rate from 1980 to 1990 was 1.4 percent; from 1990 to 1998 it was 1.1 percent.[27]

Market liberals and institutionalists both see globalization as an overall positive force for the environment because it generates the wealth necessary to pay for environmental improvements. Market liberals argue that such improvements will follow naturally from the functioning of open and free markets. States and institutions certainly have an important role to perform in terms of making and enforcing environmental policy, but it should be a minimal and market-friendly one. Globalization, by stressing liberalization of trade, investment, and finance, is also lowering inefficient trade barriers and state subsidies. This means fewer market distortions—such as prices that undervalue a natural resource. It means, too, fewer barriers to corporate investment in developing countries.

Institutionalists see a somewhat greater role for the state, and also see a need to build global-level institutions and agreements to more actively guide economic globalization (which to some extent arise naturally from the process itself). The goal is to help states, for example, advance to a higher order of development with as little environmental damage as possible. The UNEP *Global Environment Outlook* nicely summarizes the institutionalist case:

The pursuit of individual wealth on a global economic playing field made level by universal governance mechanisms to reduce market barriers can . . . open the way to a new age of affluence for all. If developing country institutions can be adapted to benefit from the new technologies and the emerging borderless economy, and if appropriate forms of global governance can be created, the rising tide of prosperity will lift everyone to new heights of well-being.[28]

Market liberals and institutionalists stress, too, the need to evaluate the environmental effects of globalization within a historical perspective. For them it is particularly important to plot and analyze global trends. Political scientist Bjørn Lomborg, an outspoken and controversial academic who is perhaps the strongest advocate of understanding the global environment in terms of global statistics, writes: "Mankind's lot has actually improved in terms of practically every measurable indicator." He warns against relying on "stories" (examples), because this can distort the analysis of progress, creating either an overly optimistic or pessimistic assessment. Lomborg continues: "Global figures summarize *all* the good stories as well as *all* the ugly ones, allowing us to evaluate how serious the over-all situation is." He further admonishes those who accept data uncritically, citing examples where scholars have come to accept a "sweeping statement" as fact when the so-called fact has no statistical base, instead resulting from "a string of articles, each slightly inaccurately referring to its predecessor" (with a far more modest original source). Lomborg argues, too, that some groups, like Greenpeace and the Worldwatch Institute, have a vested interest in painting a picture of a world in crisis. It is their job; it justifies their moral and financial existence.[29]

Look, say market liberals and institutionalists, at the world at the beginning of the twentieth century. Back then life was short and full of hardship and suffering. About a third of the global population faced possible starvation. Infectious diseases like typhoid, tuberculosis, botulism, and scarlet fever (often spread in contaminated food, milk, and water) were a leading cause of death. Global life expectancy was a mere 30 years. Even in the United States it was just 47 years, with infant mortality at a rate of 1 in 10.[30] Great progress has been made since then, not coincidentally, market liberals in particular say, as the world has become more globalized. Food production has surpassed population growth, and today less than one-fifth of the world's population is malnourished, compared to 37 percent in 1969–1970.[31] Even in the Third World the average person now consumes 38 percent more calories than in 1961.[32] It is now widely argued that famines that occur today are a result of government mismanagement, not a shortage of food.[33]

Vaccines, antibiotics, and better medical care now save millions of lives. So do refrigeration, pasteurization, and safer food-handling practices. As a result global life expectancy is far higher today—now over 66 years. Steady rises in life expectancy (see figure 2.1) have occurred in both low- and high-income countries, and there is every reason to expect these trends will continue. Economist Julian Simon sums up the case: "The standard of living has risen along with the size of the world's population since the beginning of recorded time. There is no convincing economic reason why these trends toward a better life should not continue indefinitely."[34] Market liberals believe that the process of economic globalization itself will spread higher standards of living around the world because of its ability to generate economic growth. Institutionalists see the potential for globalization to raise living standards via economic growth, but argue that global institutions that promote cooperation on these fronts have been and still are necessary to ensure that this happens.

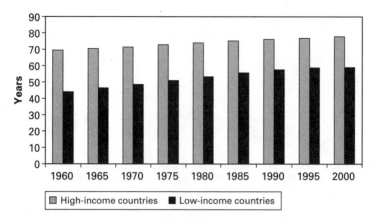

Figure 2.1
Life expectancy at birth (high-income and low-income countries). *Data source*:
World Bank World Development Indicators: www.worldbank.org/data.

Figure 2.2
World population growth. *Data source*: U.S. Census Bureau.

Progress on the food and health fronts has been accompanied by a much larger global population, and some see a direct causal link here. From 1 billion in 1804, world population had doubled by 1927 (123 years later), and jumped to 3 billion by 1960 (33 years later). By 1999 it had doubled again to 6 billion (just 39 years later) (see figure 2.2).

The global population is now well over 6 billion and is making steady progress toward 7 billion. Most analysts expect it will continue to rise into the foreseeable future, leveling off at 8–11 billion by 2050. Figure 2.3 provides three population projections for 2050: a low of 7.9 billion,

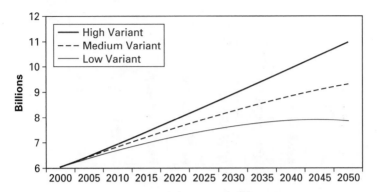

Figure 2.3
World population prospects. *Data source*: Population Division of the
Department of Economic and Social Affairs of the UN Secretariat, 2001.

a middle of 9.3 billion, and a high of 10.9 billion. (The difference is
based on alternative possibilities of future birth and mortality rates.) The
most likely scenario is the middle one. But for market liberals, even the
highest estimate shows that the global community has managed to overcome the threat of exponential population growth.

An earth with 6 billion people, market liberals point out, is not any-
where close to reaching the limits of its capacity. Abundant resources
remain. And waste sinks are far from full. Market liberals assume
humankind will be able to provide a decent standard of living for all well
into the future, provided a free and open global economy is encouraged.
Higher incomes and more modern economies allow for higher levels of
education for the general population and empower women with more
choices. This naturally lowers birth rates. The history of the world's most
advanced economies in Europe, North America, and Asia since the
Industrial Revolution some 300 years ago demonstrates this conclusively.
Institutionalists are more cautious, recognizing that there may be some
scarcities associated with population growth, but they argue that global
cooperation on improved education, economic development, and family
planning can help address the problem.[35]

For both institutionalists and market liberals, the global community
has demonstrated it can solve global environmental problems. Biotechnology has, for example, produced crops able to resist insects and diseases as well as grow in dry and inhospitable areas, and both groups see

this as having great potential for improving global food supplies. Market liberals argue that these technologies will spread via market mechanisms and benefit farmers in rich and poor countries alike.[36] Institutionalists have a bit less faith in the global marketplace to spread these benefits equally. While they see the potential of such crops to address future food needs, they argue for public funding for intergovernmental research that focuses on developing bioengineered crops that will benefit the poorest countries.[37]

Perhaps the most successful global cooperative effort was the one to reduce the amount of chlorofluorocarbons (CFCs) released into the atmosphere. CFCs were invented in 1928. Production and consumption rose quickly from the 1950s to the 1970s, mainly for use in aerosols, refrigerators, insulation, and solvents. In 1974 scientists discovered that CFCs were drifting into the atmosphere and depleting the ozone layer. This layer protects us from the harmful effects of ultraviolet sun rays, which can contribute to skin cancer and cataracts, decrease our immunity to diseases, and make plants less productive. In the decade after 1974 global negotiators worked on a collective response to this problem. This effort gained momentum in 1985 after a "hole" (really a thinning) in the ozone layer was discovered over the Antarctic.[38] In that year the global community concluded the Vienna Convention for the Protection of the Ozone Layer. The 1987 Montreal Protocol on Substances That Deplete the Ozone Layer was adopted within two years and set mandatory targets to reduce the production of CFCs. The Montreal Protocol entered into force in 1989. Amendments to strengthen the Montreal Protocol were made in London in 1990, Copenhagen in 1992, Montreal in 1997, and Beijing in 1999. The result was a significant lowering of CFC production (see figure 2.4).

The UN Environment Programme (UNEP) predicts that the ozone layer will repair itself and return to pre-1980 levels by 2050.[39] Institutionalists in particular see this as a resounding example of global cooperation to solve an environmental problem. There has been wide acceptance of the global agreements and strong compliance with them.[40] Market liberals also like this example because it demonstrates the ability of markets to respond to a global environmental problem with the development of substitutes that are less harmful.

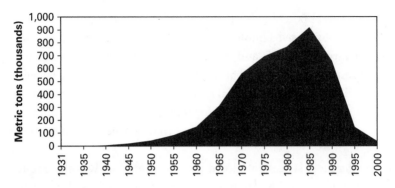

Figure 2.4
Global CFC production. *Data source*: Alternative Fluorocarbons Environmental
Acceptability Study: www.afeas.org.

Global Negative

Unlike market liberals and institutionalists, bioenvironmentalists and
social greens see economic globalization as the cause of many of the
world's environmental and social ills, rather than as its potential savior.
Both bioenvironmentalists and social greens agree with market liberals
and institutionalists that it is driving global macroeconomic growth—
but this growth is driving the overconsumption of natural resources and
the filling of waste sinks. Moreover, economic growth is not enough to
ensure well-being in a society. For bioenvironmentalists economic growth
also partly explains the exponential growth of the global population,
which for them is of primary concern. But social greens reject the bioenvironmentalists' population argument and focus instead on the way
globalization, in their view, contributes to global economic inequalities
that exacerbate environmental problems.

Bioenvironmentalists and social greens are far more critical of the so-
called progress of our so-called creations and advancements in the era
of globalization. The former president of the Nature Conservancy put
this well: "In the end, our society will be defined not only by what we
create, but by what we refuse to destroy."[41] Humans certainly live longer.
And higher incomes and better medicine and sanitation have undoubtedly made the lives of some more comfortable. Yet such data can also
hide disturbing trends. Especially worrying is the steady rise in global
cancer rates, even when adjusting rates for an aging population. Every

year 6 million people die of cancer. Ten million more are diagnosed with cancer. Only cardiovascular disease kills more people in industrialized countries. Scholars like philosopher Peter Wenz believe that the probable causes of higher cancer rates arise from the artificial changes to our living environments. There seems to be little cancer in traditional societies. One of the most significant causes of cancer in industrial societies for these scholars is the increasing volume of pesticides in the global ecosystem. In the United States alone farmers now use over a billion pounds of pesticides per year, compared with 600 million per year in the 1970s and 50 million per year in the 1940s. Cancer specialists, Wenz notes, concentrate on finding the "cure" rather than the "cause" of cancer, partly because uncovering the causes would challenge "our entire way of life."[42]

While market liberals and institutionalists tend to focus on the social and political history of the last few hundred years, bioenvironmentalists in particular like to look at the impact of globalization against the background of geological time. The universe is about 15 billion years old. Earth is about 4.6 billion years old. Modern humans have existed for just over 100,000 years and civilization for about 10,000 years. Within this time frame it is clear that humans have become a threat to the planet in a remarkably short period of geological time. Philosopher Louis Pojman provides a vivid image of this:

If we compacted the history of the Earth into a movie lasting one year, running 146 years per second, life would not appear until March, multicellular organisms not until November, dinosaurs not until December 13 (lasting until December 26), mammals not until December 15, *Homo sapiens* (our species) not until 11 minutes before midnight of December 31, and civilization not until one minute before the movie ended. Yet in a very short time, say less than 200 years, a mere 0.000002% of Earth's life, humans have become capable of seriously altering the entire biosphere. In some respects we have already altered it more profoundly than it has changed in the past *billion* years.[43]

With this outlook on time, bioenvironmentalists argue that we will soon reach the limits of the earth's biological capacity to support human life.

For bioenvironmentalists, human population growth is often the most central part of the problem—for them linked to both economic growth and globalization. They argue that a much better reflection of population patterns arises when we begin in AD 1 (see figure 2.5). Each year the global population increases by approximately another 80 million.

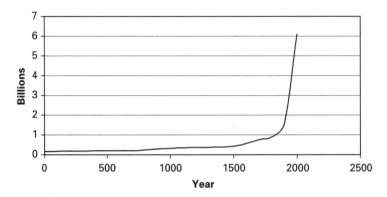

Figure 2.5
World population: AD 1–2000. *Data source*: Facing the Future: People and
Planet: www.facingthefuture.org.

Since 1950, an era of rapid globalization, the world's population has
grown more than it had in all of human history.[44] Even accepting the
midrange estimate of population growth in figure 2.3, by the middle of
this century it will exceed 9 billion—another 3 billion people to feed,
clothe, and house. Over 85 percent of these people will live in the poor
countries. Half of the population growth will occur in just six countries:
India (with 21 percent of the total increase), China, Pakistan, Nigeria,
Bangladesh, and Indonesia. Over this time the populations of the forty-
eight least developed countries will triple in size, already areas of terrible poverty and hardship.[45]

Bioenvironmentalists stress that the current rate of population growth
means the planet is increasingly awash with consumers, each demand-
ing a portion of the earth's resources. One noteworthy indicator of rising
consumption in an age of globalization is the steady increase in the
number of motor vehicles. Every year the world has absorbed another
16 million new vehicles on the road. Passenger cars now account for 15
percent of total global energy use.[46] Environmental trends confirm the
disastrous consequences of such consumption. The air in the rich industrialized countries has undoubtedly become less polluted since the 1960s,
but it is now much worse in the developing world—especially in megac-
ities like Tehran, New Delhi, Cairo, Manila, Jakarta, and Beijing. The
planet has absorbed 100,000 new chemicals since 1900.[47] Over the twentieth century total carbon dioxide emissions grew twelve times. In the

Third World more than 90 percent of sewage and 70 percent of industrial waste is still disposed of (untreated) into surface water.[48]

Global water consumption rose six times from 1900 to 1995.[49] Dams, canals, and diversions now disrupt almost 60 percent of the world's largest rivers. Deserts now threaten one-third of the world's land surface and the World Health Organization estimates that 1.1 billion people do not have access to clean water.[50] Diseases from bad water—like dysentery and cholera—kill 3 million people each year.[51] During the last century the world lost half of its wetlands. Around 30 percent of the world's coral reefs are gone and 18 percent are now at risk. Nearly 70 percent of commercial global marine fish stocks are now overfished or are at the biological limit. One-fifth of the world's freshwater fish species are extinct or endangered. Every day 68 million metric tons of topsoil washes away. Every day 30–140 species go extinct. The Worldwatch Institute, in the *State of the World 2003*, points to the dire state of global biodiversity, noting that "prominent scientists consider the world to be in the midst of the biggest wave of animal extinctions since the dinosaurs disappeared 65 million years ago."[52] The future of global biodiversity under current trends looks bleak. A study by nineteen scientists published in the journal *Nature* in 2004 predicts that a temperature rise from global warming of 0.8°–2.0°C will "commit" 18–35 percent of animal and plant species "to extinction" by the middle of the twenty-first century.[53] Other factors attributed to global warming—such as higher concentrations of carbon dioxide—could mean an even higher rate of extinctions.[54]

The history of deforestation is typical of the destructive impact of the global political economy. Almost half of world's forests are now gone. Between 1980 and 1995 alone, over 180 million hectares of forest were lost, mostly in developing countries. That is six times the size of the Philippines, ten times Cambodia, and sixty-four times the Solomon Islands. Frontier forests (defined by the World Resources Institute as large and pristine enough to still retain full biodiversity) are under some of the greatest pressures. Asia, for example, has lost almost 95 percent. The Philippines no longer has frontier forests. Vietnam, Laos, Thailand, and Burma will soon follow. Cambodia has just 10 percent, Malaysia just 15 percent, Indonesia only 25 percent, and Papua New Guinea only 40 percent.[55]

Social greens, too, point to the steady decline in the state of the global environment in an age of globalization. They agree strongly with bioenvironmentalists that economic growth and overconsumption are serious threats to the planet. Yet unlike bioenvironmentalists, they do not focus on population growth as a primary driver of such problems. They point out that growth in consumption of forest products, food, and water, for example, far exceeded population growth over the past thirty years.[56] Instead, they place their main focus on global inequality and its associated environmental problems—those linked to both overconsumption among the wealthy and the dispossession of the poor from their traditional lands.

For social greens, economic globalization is the primary cause of this inequality.[57] It is seen as reinforcing neocolonial relationships between rich and poor countries, as well as changing production patterns in complex ways that have serious environmental implications.[58] The International Forum on Globalization, an international group of activists and academics opposed to globalization, points out that today's world is "in a crisis of such magnitude that it threatens the fabric of civilization and the survival of the species—a world of rapidly growing inequality, erosion of relationships of trust and caring, and failing planetary life support systems."[59] Life may be easier—and perhaps even better—in sections of the wealthier countries (although not in the slums). But in much of the developing world it is worse. Since 1972 the number of people living in extreme poverty (less than US$1 a day) has grown to 1.2 billion people. Almost half the world's population (2.8 billion) survives on less than US$2 per day.

There is not enough food in many countries. In fact, even though global food production is now higher, in many parts of the world, such as in Africa, food production has lagged behind population growth over the last two decades.[60] In India, where one-third of the 1 billion people live in poverty, over half of the children are undernourished. Worldwide, over 840 million people suffer from chronic malnutrition, which contributes to 60 percent of all childhood deaths.[61] And nearly 11 million children die unnecessarily every year (that is, from preventable and treatable causes). Of these, 8 million are babies, half of whom die in the first month of life.[62] At the same time, obesity rates are rising fast in the world's wealthy countries—in the United States, for example, it went from 12 percent of the population in 1991 to 17.9 percent in 1998. The largest rises in obesity are among 18- to 29-year-olds.[63] In 1999 over 60 percent of all adults in the United States were overweight. Obesity and physical inactivity now account for over 300,000 premature deaths annually in the United States alone. In late 2001 the U.S. Surgeon General issued a "call to action" to address the crisis of obesity, claiming "overweight and obesity may soon cause as much preventable disease and death as cigarette smoking."[64]

Great inequality is also seen in the distribution of disease. Infectious diseases remain the number one cause of death worldwide, accounting for roughly a quarter of total global deaths. Over 13 million people die each year from diseases like AIDS, malaria, tuberculosis, cholera, measles, and respiratory diseases, most of these in the developing world. These high rates of death from such diseases only contribute to further poverty. In some African countries, life expectancy has fallen, mostly because of AIDS, malaria, and war. The AIDS pandemic started two decades ago. Now over 60 million people are infected with HIV, with numbers rising rapidly especially among the young (half of all new HIV infections are adolescents). In 2001 AIDS killed 3 million. The UN predicts that 68 million more will die prematurely of AIDS by 2020. Of these, 55 million will live in sub-Saharan Africa. Already AIDS has lowered the average life expectancy in this region from 62 to 47 years. In Botswana, Namibia, and Zimbabwe, life expectancy has dropped from 60–70 to 38–43 years. In Tanzania, 40 percent of the population is now dying before the age of 40. Across Africa, there are about 5,500 funerals a day for AIDS victims. UN AIDS has predicted that one in every five adults in Africa will die of AIDS in the next ten years, destroying the social fabric of much of Africa.[65] The death rates from AIDS are much higher in the developing world because of that region's lack of access to the latest drugs and treatments that are currently available in the rich industrialized countries.

Social greens argue that this represents unnecessary suffering on the part of the majority of the planet's inhabitants. The inequality that has accompanied economic globalization is the driver not just of poor social conditions for many, but also of environmental problems. The poor are put in situations where they are uprooted from their lands and forced to degrade what lands they can just to make ends meet. And the rich, meanwhile, overconsume and contribute to a worsening of industrial pollution. For these thinkers, it is not surprising that pollution levels may be falling in rich countries, because globalization has enabled the types of production that pollute most to migrate to developing countries, in the form of foreign direct investment. The rising levels of pollution in the rapidly industrializing developing countries of Southeast Asia and Latin America are evidence of this, according to social greens.[66] This transfer of environmental degradation from rich to poor countries has direct and negative impacts on the lives of the world's poorest people.

The technological solutions offered up by Western science are not a potential savior for bioenvironmentalists and social greens. The technological transfers of globalization are in fact often a deceptive solution, the heroin of market liberals and institutionalists, temporarily allowing societies to deflect a problem into the future or into another ecosystem. They also have much less faith than institutionalists in the ability of global institutions to guide globalization to ensure that it does not cause further environmental problems. Social greens in particular worry that global organizations and agreements can become props of capitalism, as is the case with the World Bank, International Monetary Fund (IMF), and World Trade Organization (WTO).

Agricultural biotechnology is just one example where social greens and bioenvironmentalists are skeptical of Western technology and institutions. Critics of genetically altered crops argue that they pose enormous ecological and health risks—from genetic pollution and erosion, to potential allergic reactions.[67] These crops are also seen as reinforcing inequalities and as giving power to the transnational corporations that patent them.[68] Global institutions, including the World Bank and the WTO, have continued to promote these technologies as beneficial, despite growing concern over their use. Many social greens, however, still see a need for global institutions, just not the ones we currently have. The International Forum on Globalization summarizes this position succinctly:

There is certainly a need for international institutions to facilitate cooperative exchange and the working through of inevitable competing national interests toward solutions to global problems. These institutions must, however, be transparent and democratic and support the rights of people, communities and nations to self-determination. The World Bank, the IMF, and the World Trade Organization violate each of these conditions to such an extent that [we] recommend that they be decommissioned and new institutions built under the authority of a strengthened and reformed United Nations.[69]

Bioenvironmentalists and social greens also have less confidence in the value of international regimes to slow or resolve global environmental problems than other thinkers do. Certainly most would agree that the global effort to reduce the production of CFCs was successful. But this was an exceptional case, one that tells us little about our ability to handle future global environmental crises. The causes and consequences of the depletion of the ozone layer were relatively straightforward. Skin cancer, one of the most visible consequences of less ozone, was a particular worry. In the mid-1980s CFCs were produced by only twenty-one firms in sixteen countries, and developed countries were responsible for about 88 percent of production. Especially important, by 1988 the chemical company DuPont, which accounted for one-quarter of global CFC production, had found an affordable substitute to CFCs. The economic cost of phasing out CFCs was therefore minimal.[70]

Most global environmental problems, say bioenvironmentalists and social greens, involve far greater complexities and uncertainties than the ozone case and will require far greater sacrifices to solve. Take global warming. The UNEP reports that the mean global surface temperature has risen by 0.3°–0.6°C in the last 100 years. This may not seem like much. But it was the largest increase of any century over the last millennium. The problem appears to be getting worse, because the 1990s was the warmest decade and 1998 the warmest year.[71] For bioenvironmentalists, trends like the one in figure 2.6 indicate our failure to act.

Global warming especially alarms bioenvironmentalists and social greens, because the three main greenhouse gases (carbon dioxide, methane, and nitrous oxide) arise from core economic activities (automobile use, electricity generation, factories, agriculture, and deforestation), while the main consequences (rising seas, severe storms, drought, and desertification) are beyond the lifetimes of politicians and business leaders—perhaps occurring in 50 to 100 years. And the impacts, when they are most severe, will be mostly felt by the poor, marginalized peoples of the world.[72] Obviously, lowering greenhouse gas emissions will involve major changes to global economic production and consumption patterns. It will require, too, governmental, corporate, and personal sacrifices.

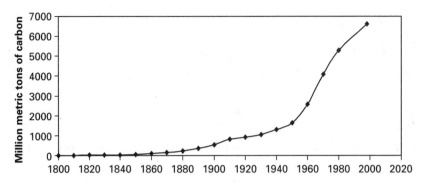

Figure 2.6
Global CO_2 emissions from fossil-fuel burning, cement manufacture, and gas flare. *Data source*: Marland, Boden, and Andres 2001.

Substituting CFCs, say these thinkers, is just not a comparable sacrifice and it is overly naive to believe that it demonstrates that the global community has the collective capacity to act quickly, coherently, and effectively whenever a problem is deemed urgent enough. Any progress on climate change agreements, rather, is more likely linked in some way to corporate strategies to break into renewable energy markets.[73]

Conclusion

This chapter has shown the power of globalization as a force of economic and, at a slower rate, societal and political change. Each worldview brings its own lens to the process, illuminating different aspects—some more positive, others more negative. Each view thus brings different insights into the implications of globalization for the environment.

The market-liberal lens focuses on the benefits of growth and the power of free markets to foster it. For market liberals globalization is a force of good, an engine of progress. It promotes efficient production and trade of goods, diffusing appropriate technologies to areas with natural labor and resource advantages. It promotes investment that brings in environmental technologies, critical funds, and better management. This advances the economies toward less destructive activities by diversifying the sources of economic growth, which in the long run allows economies to shift away from a heavy reliance on natural resource exports. Globalization also means more macroeconomic growth, raising per capita incomes throughout the world. The higher national per capita incomes that arise from embracing globalization are creating the societal and political will to tackle national, and ultimately, global environmental problems. In the long run economic growth extends lives, improves health care, and raises national environmental standards. Globalization allows for even more wealth. Such wealth will allow for even more economies to shift away from agricultural and industrial production and toward knowledge-based economies.

The institutionalist lens is sympathetic to the points raised by market liberals, but focuses more specifically on the international cooperation necessary to bring those benefits to fruition. Institutionalists agree that economic growth and free markets can have enormous benefits for the environment. Yet they argue that globalization should not be seen as a panacea. The process, like any dynamic force, is producing *both* positive and negative changes for the global environment. It is good in macroeconomic terms. It is raising global wealth; it is fostering innovation. But the rewards are not always equitable. At times, globalization seems to aggravate pockets of environmental degradation and inequality. From an institutionalist perspective, it is therefore best to guide globalization—to use institutions and cooperation and intelligence to maximize the economic benefits and minimize the social inequities. This requires strong national and local governments as well as powerful global organizations. The communication technologies of globalization are advancing this cause, by integrating cultures, fostering awareness, and creating global norms, standards, and sympathies that encourage cooperative efforts to address national problems like malaria or global problems like climate change. The globalization of norms, codes of conduct, environmental markets, environmental organizations, and international law is deepening global environmental governance, adding another layer of control above national-government regulations.

Bioenvironmentalists look at globalization through a very different lens. Instead of seeing opportunity in globalization, they focus on the scarcity it exacerbates. For these thinkers the future is bleak. Globalization props up unsustainable population growth in developing countries, as the rich drop food and medical supplies into poor countries with population growth rates embedded in a culture of poverty. Globalization is also herding populations into overcrowded cities or fragile ecosystems (like onto land cleared of its rainforests), and turning rural areas into vast and specialized farms to feed the cities, often in distant lands. At the same time, the globalization of trade, investment, and financing is accelerating global economic growth. This creates more output and more consumption. It creates, too, more opportunities for the rich to over-consume and waste resources, deflecting the ecological impacts of this consumption overseas or into the global commons. Globalization here is little more than "eco-apartheid."[74] It is deepening the global culture of consumption for consumption's sake. People are losing the sense of "enough," as the rising rates of obesity in the developed world show. Ingenuity and technology can certainly help to mitigate particular problems. Yet technology cannot solve the ecological consequences of globalization. Solutions can only begin to occur once the human species accepts that it is now beyond its carrying capacity.

Social greens agree with much of the bioenvironmentalist analysis of the ecological consequences of globalization. Yet their lens focuses much more on the social injustice arising from globalization than on scarcity per se. Social greens see globalization as the driving force behind the spread of global inequality and large-scale capitalism and industrial life, which for them are core causes and consequences of the global ecological crisis. Inequality and the imposition of industrialism contribute to the eradication of the rights of indigenous peoples, women, and the poor, and to the destruction of culture as capitalists reconstruct societies into new markets and production nodes. These forces are eroding the autonomy of communities and creating a consumer monoculture. Globalization, for many social greens, is little more than ecoimperialism, a process to siphon off autonomy and knowledge from the local to the global. In this way globalization is reinforcing patterns of economic, environmental, and social injustice. The globalization of production and trade, more-over, further distances an individual's ability to perceive the ecological and social impacts of these behaviors. People are increasingly unable to see (or at least are able to forget) how their everyday choices damage the environment or injure workers.

The four worldviews, in short, have markedly different interpretations of the ecological effects of globalization. Each offers its own unique insights by focusing on particular aspects of the process. The rest of this book will explore the debates between these worldviews in far greater depth, drawing out subtleties of opinion both within and across the worldviews. The rather stark characterization of worldviews on globalization in this chapter will, we hope, help to guide you through the thicket of the more specific, and at times overlapping, arguments in the remaining chapters. We begin in the next chapter with what all four worldviews agree is one of the most significant effects of globalization—the global spread of environmental norms, ideas, and institutions over the past half century.

3

The Globalization of Environmentalism

What has the global community done to tackle the environmental problems discussed in the previous chapter? While there is a rich history of formal actions on the part of states to address these problems in the international arena, we must also remember that the history of global environmental politics is inextricably tied to contests of ideas: battles of worldviews and discourses.[1] We have seen new environmental ideas and language enter into mainstream discourse as global awareness rises and as environmental conditions deteriorate. These views are also affected by wider developments in the global political economy. Accounts of the history of global environmental politics, particularly those in the field of international relations, have focused on environmental diplomacy. Such studies tend to explain the evolution of environmental discourse primarily in terms of the outcomes of environmental summits, commissions, agreements, and organizations.[2] These political-diplomatic efforts have certainly been important. But a deeper look at this history reveals that parallel developments in the global political economy have, directly and indirectly, often set the wider agendas for these formal discussions, affecting the evolution of environmental ideas. Tracing this history of both environmental diplomacy and the global economy, then, helps uncover when and why the various arguments of bioenvironmentalists, institutionalists, market liberals, and social greens first emerged. It also helps us to better understand the roles of various actors today in global environmental governance.

The first part of this chapter provides a brief history of the globalization of environmentalism over the past half century, highlighting the connections of this diplomatic, interstate history to events in the global economy and to the emergence of different perspectives on the global environment. The second part of the chapter outlines the various mechanisms of global environmental governance that have been put into place over this period, and highlights the significance of each in today's global environmental discourse.

The Evolution of Global Discourse on Environment and Development

Homo sapiens seem to have had only a marginal environmental impact until relatively recently (from perhaps 500,000 years ago to 10,000 years ago). The global population was low and fairly stable, and the main use of natural resources was for stone, bone, and wooden tools. The shift from hunting and gathering to settled agriculture that began about 10,000 years ago was a profound turning point in global environmental history. The global population at this time was probably a mere 10 million or so. The food surpluses from settled agriculture helped populations expand into civilizations (cities). These civilizations, with more elaborate divisions of labor, in turn became great sources of technological advances. The plow (animal drawn) and the wheel as well as writing systems and the use of numbers were invented during this time. These civilizations were also able to extract and use larger and larger amounts of natural resources—like metals and wood. Not all of them, however, could manage their natural environments in the long term. For example, the collapse over 4,000 years ago of Mesopotamia, a land between the Tigris and Euphrates rivers, was connected to an irrigation system that poisoned the soil with salt.

Yet it was not until some 300 years ago, with the dawn of the Industrial Revolution, that human activities began to accelerate toward a scale able to alter the global environment. The worsening impact of the environmental consequences of industrialization—such as the burning of coal—along with rising populations partly explains the emerging concern with environmental conditions over the last several centuries. The first of a series of killer fogs (air pollution from coal burning) struck London in 1873, killing more than 1,150 people. Overhunting and habitat loss brought the plains bison of North America to near extinction by the mid-1800s. The last passenger pigeon died in a zoo in 1914, a bird that once migrated through eastern North America in the millions.[3] Poor land management was seen as a factor behind the "Dust Bowl" conditions (drought, dust storms, and agricultural collapse) in Western Canada and the United States in the 1930s. By 1950 the world population was over 2.5 billion. The effects of industrialization on the health of these growing populations were becoming an increasing concern. Smog in London killed 4,000 people in 1952. A year later 200 died from smog in New York City. In 1956 scientists officially "discovered" the mercury poisoning of villagers in Minamata, Japan (commonly called the Minamata disease). The poison came from the local seafood, which had absorbed methyl mercury dumped into the sea by the Chisso industrial plant. Hundreds eventually died, and hundreds more were born with brain damage and other birth defects.[4]

The emergence of environmentalism from the 1700s to the mid-1900s, then, can be traced in part to environmental turmoil within the industrialized world. It was also, however, tied to the experience of the colonial powers at the time. Extensive economic and political control characterized European colonies in Latin America, Africa, and Asia. Other countries, such as the Untied States, exerted economic if not political control in some of the same parts of the world—what some label economic imperialism. Colonialism and imperialism had enormous environmental consequences, especially as many previously unconnected areas of the globe became primary sources of raw materials for Western industrialization. Much of this environmental damage is still evident today.[5] The colonies were also the testing grounds for many Western environmental ideas, and even today the results of these "tests" continue to shape global environmental discourse. The primary environmental goal of many colonial governments was to preserve and better manage nature and wildlife to maximize its aesthetic and economic value. The conservationist ideas of this era were certainly "Western" in their origin, and often administrators from countries like Britain, France, and the United States simply imposed these views on their colonies. In other cases environmental ideas first emerged within the colonies as colonial administrators attempted to "better" manage resource extraction and economic development. Only later were these ideas brought back to the imperial centers.[6]

Many colonists were horrified at how indigenous peoples treated nature, believing them ignorant and wasteful. The British and French, for example, believing indigenous slash-and-burn agriculture (swidden farming) was primitive, put in place new bureaucracies to manage forests.[7] The purpose of these bureaucracies was to ensure the colonies managed the natural environment "efficiently" and "rationally." Colonial administrators became concerned too with hunting in Africa during the nineteenth century, as the continent was stripped of much of its wildlife. In response legislation was enacted to create parks and game preserves as well as to regulate hunting. The latter was typically denied to Africans on the grounds that sport hunting was less destructive than that for livelihood.[8]

The development and diffusion of Western environmental ideas within the colonies and from the colonies to the imperial countries can be seen as a first step in the creation of a more "global" environmental discourse. Often, colonizers attempted to back these views with the language of science, developing models like sustainable yield management, with the purported goal of managing resources seen as fully renewable. Many scholars have since dubbed this "pseudoscience," however, arguing that these views were paternalistic, ideological, and racist.[9]

Efforts to conserve nature in both the colonies and in the metropole countries continued into the twentieth century, though the two world wars and the Great Depression of the 1930s hampered global coordination. The newly formed United Nations took on such efforts shortly after the end of the Second World War, contributing to the creation of the International Union for the Protection of Nature in 1948, which was renamed the International Union for Conservation of Nature and Natural Resources (IUCN) in 1956.[10] This organization, which today comprises 70 states, 100 government agencies, and over 750 non-governmental organizations (NGOs), was designed to coordinate global efforts to conserve and preserve nature.[11]

Silent Spring and the 1960s and Early 1970s

The global economy boomed for nearly 20 years after World War II, ultimately fueling the environmental movement. Concern with the environmental abuses of rapid industrialization gained momentum in the First World in the 1960s. There were protests against nuclear weapons and chemical pollution. The World Wildlife Fund (known today as the WWF) was founded in 1961 to work to preserve global biodiversity. Rachel Carson's bestseller *Silent Spring* (1962) was particularly influential with the public. Her message was a powerful one: the increasing use of chemicals, particularly pesticides, was killing nature and wildlife. It told of a future spring without the songs of birds. The scientific community lost no time in attacking Carson in the media, accusing her and her supporters of overstating the case against DDT (dichlorodiphenyl-trichloroethane) and other chemicals.[12] But the strength and simplicity of her message endured with the general public, and indeed research has since shown that her concern over DDT and other synthetic chemicals, many of which are now banned from production and use, was warranted.

The need to protect nature (so, for example, people could enjoy the "outdoors") remained a significant concern for the environmental movement in the First World. But increasing public concern with the effects of industrial production began to give the movement a new focus. Also emerging were worries among environmentalists about the cumulative impact of local problems on the health of the planet. More and more people began to see the planet as fragile and interconnected, an image reinforced as pictures of earth from space became more common. For many, the most memorable picture was the one that became available as astronauts Neil Armstrong, Michael Collins, and Edwin Aldrin journeyed to the moon in July 1969 (see box 3.1). Even today, the image of the earth from space—Spaceship Earth with no visible borders and a common atmosphere and oceans—remains a powerful symbol for global environmental consciousness.[13]

Other developments shaped the discourse as well in equally profound ways. The early 1960s saw rapid economic growth and greater global integration of trade and investment. Global growth rates were as high as 5–6 percent in the early to mid-1960s (though this growth was not evenly distributed among countries or regions).[14] The global economic infrastructure codified by the International Monetary Fund (IMF), the World Bank, and the General Agreement on Tariffs and Trade (GATT), all set up at the end of World War II, was widely seen as critical to fostering this growth and integration. Advances in communication and transportation technologies also contributed to the economic growth in this period. The wave of decolonization in Asia and Africa from the late 1940s to the late 1960s put many newly independent colonies on the global stage. Most joined the global economic institutions in an effort to

Box 3.1
The earth from space

Photograph of the earth on July 16, 1969, taken from Apollo 11 at 180,000 kilometers. It shows most of Africa and the Middle East as well as parts of Europe and western Asia.
Source: NASA photo ID AS11-36-5355.

promote economic growth. It was a time of optimism, and many countries believed that development (often understood as rapid industrialization and economic growth) would follow automatically from participation in the global economy.

Yet by the late 1960s and early 1970s many of these "developing" countries realized that replicating the industrialization in the First World was far from automatic. Global growth rates had been strong, but the Third World still lagged far behind the First World in terms of industrialization and industrial exports.[15] Critics from developing countries began to argue that the postwar global economic infrastructure still reflected colonial and imperial interests. Dependency theorists argued that the imperial nature of capitalist relations between rich and poor countries made the kind of industrial development seen in the capitalist North difficult, if not impossible, to replicate in the South.[16] The rich countries were seen to be dependent on the developing countries for cheap supplies of raw materials, while the elites in developing countries were dependent on capitalist patrons in the rich countries. Other critics also argued that it was particularly difficult to break out of a cycle of being mere suppliers of raw materials for overseas production, mainly because of the difficulty of earning enough money from those exports to foster industrialization. This was due to a serious problem faced by developing countries, referred to as "declining terms of trade," which occurs when the prices for exports (in this case raw commodities) are generally low or falling, while prices for imports (in this case industrial goods) are generally high and rising. Under these conditions earning enough from raw material exports to purchase the capital equipment necessary to industrialize is extremely difficult. This focus on unequal terms for rich and poor countries in the global economy has been a longstanding perspective in international political economy, and has been important in informing the views of the social greens and bioenvironmentalists.

Views critical of the inherent inequality of the global economy spread as global growth rates began to falter in the early 1970s. Rising inflation and a U.S. trade deficit spurred President Richard Nixon in 1971 to take the U.S. dollar off the gold exchange standard established at Bretton Woods in 1944. This action only further undermined confidence in the global economy.[17] This precarious global economic situation added to the growing sense of global interconnectedness, as well as mutual vulnerability. Many environmentalists in the First World began to argue that because environmental problems and economic performance were intertwined, a single global economy was surely tied to the fate of a single global environment.

Environmentalism in rich countries at this time began to reflect the economic theme of mutual (North-South) vulnerability. Books like Paul Ehrlich's *The Population Bomb* (1968) became popular bestsellers. Ehrlich argued that population growth in the Third World, spurred in part by its economic predicament, would one day threaten the globe's resource base. Inequalities would continue to grow in such a world, although both the rich and the poor would eventually suffer from mass starvation, violence, and ecological destruction. He warned: "It is obvious that we cannot exist unaffected by the fate of our fellows on the other end of the good ship Earth. If their end of the ship sinks, we shall at the very least have to put up with the spectacle of their drowning and listen to their screams."[18] The book sold 3 million copies and went through 22 printings in the 3 years following its release. Coming from a scientific framework and focusing on unsustainable population levels, Ehrlich was one of the key thinkers in the emerging bioenvironmentalist perspective, which began to gain popularity in industrialized countries.

Environmental concern continued to rise in the North in the late 1960s and early 1970s. On November 30, 1969, Gladwin Hill of the *New York Times* reported on the situation in the United States: "Rising concern about the environmental crisis is sweeping the nation's campuses with an intensity that may be on its way to eclipsing student discontent over the war in Vietnam."[19] Twenty million rallied in the United States at the first Earth Day on April 22, 1970, one of the biggest organized demonstrations in the history of the United States. In that year the U.S. government established the Environmental Protection Agency (EPA). Canada followed a year later with a Department of the Environment. The Canadian group Greenpeace made headlines in the same year when it sailed toward Amchitka Island, Alaska, to protest against underground nuclear tests.

Feeding into this rising interest in environmentalism in the North was another bestselling and controversial book, *The Limits to Growth*.[20] This book reported in more popular language the findings of a technical study by a research team at the Massachusetts Institute of Technology. It was part of a wider project of the Club of Rome, an international research and policy group. Its conclusions were similar to Ehrlich's. Using formal modeling and computers, it simulated future prospects for five major, interconnected trends: industrialization, human population growth, natural resource depletion, malnutrition, and pollution. It predicted that if these trends continued, the earth would reach its limits within 100 years. If allowed, the book argued, "The most probable result will be a sudden and uncontrollable decline in both population and industrial capacity."[21] But the authors did not forecast inevitable doom. They felt it was possible to alter the growth trends to foster sustainability, both ecologically and economically. In making such arguments, the book was a key outline of the early bioenvironmentalist views. It sold 12 million copies and was translated into 37 languages. There were many attacks on both the work of Ehrlich and the Club of Rome for what were seen by many to be exaggerations.[22] Donella Meadows, one of the original authors of *Limits to Growth*, on reflection, saw such virulent attacks on her work as inevitable given the dominance of the economic growth paradigm. As she notes, "A book called *The Limits to Growth* could have been filled with nothing but blank pages, and some people would still have thought it was an anticapitalist plot or an anti-Communist or an anti-Keynesian or an anti–Third World one. They never saw the book for its cover."[23]

The following year saw the publication of another extraordinarily influential book in the West: E. F. Schumacher's *Small Is Beautiful: Economics as if People Mattered*. This book helped to shape the development of social green thought with its focus on the need for small, local economies. Schumacher was deeply influenced by Indian anticolonial activist Mahatma Ghandi, whose nonviolent means of protest in the lead-up to Indian independence in 1947 won him international acclaim. Ghandi also outlined a vision for a free India organized around self-reliant and self-governing villages, which encouraged local household-level production, rather than large-scale industrial society.[24] Schumacher was also influenced by Buddhism, arguing that a traditional economist

is used to measuring the "standard of living" by the amount of annual consumption, assuming all the time that a man who consumes more is "better off" than a man who consumes less. A Buddhist economist would consider this approach excessively irrational: since consumption is merely a means to human well-being, the aim should be to obtain the maximum of well-being with the minimum of consumption. . . . The less toil there is, the more time and strength is left for artistic creativity. Modern economics, on the other hand, considers consumption to be the sole end and purpose of all economic activity.[25]

In making these arguments, Schumacher also helped to build the foundation of the field of ecological economics.

These views on population, consumption, and limits to growth in the 1960s and 1970s provided a foundation for the bioenvironmentalist perspective and have been a strong continuing influence in the environmental movement in the North. Though the ecological collapse predicted by many early bioenvironmentalists did not occur, those in that school today, like Lester Brown (president and senior researcher, Earth Policy Institute) and Professor William Rees, continue to argue that population pressures and limits to growth do exist, and do imperil the future of the planet.[26] The bioenvironmentalist view of the ecological interdependence of national economies strengthened the support for global institutions and agreements as the only viable response to emerging global environmental problems in a highly interdependent world.

Social green thought also began to emerge at this time. Though social greens agreed with much of the bioenvironmentalists' arguments regarding consumption and growth, they were critical of the focus on population growth in the South as a cause of environmental harm. This gave their views more appeal to environmentalists in the developing world. Social greens also felt that local solutions were superior to global ones, drawing on the inequality arguments of dependency thinkers as well as on the visions for a local economy of Ghandi and later Schumacher. The next section examines global environmental cooperation and institutions since the early 1970s.

The Stockholm Conference and the 1970s

Debates over the links between the global economy, population growth, and environmental change on the one hand, and the sense of mutual interdependence and vulnerability on the other, were at the core of the United Nations Conference on the Human Environment, held in Stockholm, Sweden, in June 1972. This was the first global-level UN conference on the environment as well as the first world conference on a single issue. Delegates from 113 governments attended. Canada's Maurice Strong was chair (see box 3.2). Three parallel NGO conferences—the Environment Forum (with official UN support), the Peoples Forum, and Dai Dong—took place alongside the official proceedings.

The conference agenda and outcomes must be seen against the backdrop of the Vietnam War and the Cold War. The United States printed more and more dollars to finance the Vietnam War, contributing to a rise in global inflation and overall global economic concerns. Many conference participants feared, given the state of the global economy, that their governments could not afford the requirements for global environmental protection. The Cold War also directly altered the political climate of the Stockholm Conference. Russia and the Eastern Bloc countries boycotted the conference because East Germany, not yet a member of the UN, was not allowed to participate. This was a blow for Strong, who was keen to ensure the participation of all countries.

Box 3.2
Maurice Strong

Maurice Strong played an important personal role in the globalization of environmentalism. Born in Canada in 1929, he started as an entrepreneur and businessman, working his way up to president of the Power Corporation of Canada. In 1966 he became head of Canada's External Aid Office (later, the Canadian International Development Agency). From 1970 to 1972 he was secretary-general of the UN Conference on the Human Environment and became the first executive director of the UN Environment Programme in 1973. He returned to Canada to become president of Petro-Canada (1976–1978). In the 1980s he was chair of the International Energy Development Corporation and the Canada Development Investment Corporation. In the 1980s he also served as a member of the World Commission on Environment and Development. After that he became secretary-general of the 1992 UN Conference on Environment and Development. In 1992 he again returned to Canada, this time as chair and CEO of Ontario Hydro, North America's largest utility company. In 1995 he was named senior advisor to the president of the World Bank, and is currently a senior advisor to UN Secretary-General Kofi Annan on UN reforms.

Source: Strong 2000.

The initial purpose of the conference, announced in 1968 by the UN General Assembly, was to discuss "problems of the human environment" and "identify those aspects of it that can only, or best be solved through international cooperation and agreement."[27] The focus was on environmental problems arising from industrialization, of particular concern to Northern governments. This was expanded, however, to include broader development concerns to gain the support of developing countries. The UN convened a meeting in 1971 in Founex, Switzerland, which brought together development experts and scientists from the developing world to discuss the links between environment and development. The *Founex Report on Development and Environment*, which is credited with encouraging wide participation at Stockholm among the developing countries, emphasized the environmental problems linked to poverty in the developing world, arguing that development, including industrialization, was necessary to overcome poverty-related environmental degradation.[28]

The focus of many of the debates at Stockholm, then, was on how to reconcile the economic development demanded by the South with the perceived need of the North to protect the global environment.[29] The phrase "the pollution of poverty" was coined at Stockholm as developing countries, reiterating some of the core conclusions of the Founex Report, argued that the greatest environmental threat was poverty. At the same time developing countries were skeptical of the North's environmental agenda, worrying it would deny poorer countries the benefits of economic growth and industrialization. Brazil, for example, declared that it had little interest in discussing industrial pollution, which it saw as a "rich man's" problem.[30] The representative from the Ivory Coast went as far as to state that the country would welcome more pollution if this would allow it to industrialize.[31] Developing countries further argued that exploitation by global capitalists was a core reason for their high levels of poverty in the first place. Global economic institutions were singled out for pushing them to export raw materials on declining terms of trade. Many developing countries called for global economic reforms as part of efforts to solve global environmental problems.

Official conference documents at Stockholm, however, did not reflect these calls. At most these documents acknowledged the unique problems of developing countries, but no real remedies were offered.[32] The conference produced a Declaration on the Human Environment, which contained 26 principles, an Action Plan for the Human Environment, which included 109 recommendations, and a Resolution on Institutional and Financial Arrangements. These measures were "soft" international law—that is, they did not legally bind the signatory states—yet they did signal a growing concern among national governments over the global environment. The Stockholm Conference is also credited with the creation of the United Nations Environment Programme (UNEP). Officially launched in 1973, UNEP was headquartered in Nairobi, Kenya, and Maurice Strong was named its first executive director.

The Stockholm outcomes, in particular the UNEP and the demand for international environmental cooperation, bolstered the views of the institutionalists. Bioenvironmentalists saw the need for legislation to control individual behavior, and argued for states to rein in both economic and population growth. Institutionalists, on the other hand, argued for international legal controls on states to address particular environmental problems. This view gained considerable legitimacy in the years following Stockholm, reflected in more and more governments entering into an increasing number of negotiations for "hard law," or environmental treaties, which are binding agreements.

The world economy spun into further turmoil in the years shortly after the Stockholm Conference. Restrictions on oil supplies by the Organization of the Petroleum Exporting Countries (OPEC) in 1973–1974, which led to a quadrupling of oil prices, spurred high inflation and low growth rates worldwide. This then filtered through to prices for all industrial goods produced with energy. Many developing countries were upset with the impact of this global economic downturn on their levels of debt and their prospects for industrial development. In 1974 the developing countries called on the United Nations to create a New International Economic Order (NIEO). They demanded reforms to counter declining terms of trade for raw commodity producers, increase voting power for developing countries in the IMF and World Bank, put controls on transnational corporations, and relieve foreign debt. A NIEO was at the center of global negotiations throughout the 1970s, but concrete measures did not get far.[33] Nonetheless, these global negotiations reinforced the sense of global economic interdependence in both the North and South.

The economic turbulence of the 1970s relegated environmental issues to the back burner in many countries, especially poorer ones. Yet states still negotiated important global environmental treaties in this period, partly as a result of processes set in motion by the Stockholm Conference. These include the Convention on the Prevention of Marine Pollution by Dumping of Wastes and other Matter (the London Convention, 1972), the Convention on International Trade in Endangered Species of Wild Flora and Fauna (CITES, 1973), and the Convention for the Prevention of Pollution by Ships (MARPOL, 1973).

Alongside these international diplomatic efforts, there were a growing number of local environmental movements to protect the environment, particularly in non-Western countries.[34] Many of these movements stressed the need to address poverty and inequality as a primary way to improve environmental conditions, seeing livelihoods and the environment as inextricably linked. This perspective gave many of these movements philosophical ties with the social greens. An example was the Chipko movement, which emerged in the Indian state of Uttar Pradesh in the Himalayas in the early 1970s. This movement was a response to deforestation linked to increased logging and industrialization, which stripped local peoples of their livelihoods, leading not only to more environmental damage, but also to higher levels of poverty. The movement, drawing inspiration from the Ghandian philosophy of promoting local and small-scale production, sought to resist government allocation of trees to large industries that denied the poor access to those resources. This grassroots movement applied direct-action methods, including tree hugging and protests, with the aim of protecting the forests for small-scale use by locals.[35] The Chipko movement was comparatively successful, gaining international recognition and spreading the methods of resistance to other parts of India.

In an attempt to ensure that states did not forgo environmental management in hard economic times, the IUCN, along with UNEP and the WWF, issued a joint report on the global environment in 1980, the *World Conservation Strategy*.[36] The report, a product of three years of collaborative work among these organizations, reflected the Northern environmental agenda of the 1960s and 1970s (drawing on elements of both bioenvironmentalist and institutionalist ideas). It focused on conservation and resource management on the one hand, and the impact of population pressures on those resources on the other. The *Strategy* outlined the goals and targets the world should aim for to preserve ecological processes, genetic diversity, and species and ecosystems. Importantly, it brought the concept of "sustainability" to the world's attention. It came close to integrating development and environment, although the emphasis was still more on the environment side. Many would later criticize the *Strategy* for failing to address the necessary changes—economic, social, and political—to achieve its ambitious goals. The concept of sustainability would, however, shape global discussions on environment and development in the 1980s.

The Brundtland Report and the 1980s
The 1980s brought more turmoil for Third World economies. A foreign debt crisis, in which the South is still mired, was first openly acknowl-

edged in August 1982 when Mexico announced it could not repay its international debts. While other developing countries followed Mexico into a debt crisis in the early 1980s, First World economies began to recover. This relatively strong performance in the North bolstered the global spread of the values and neoliberal economic ideologies of Prime Minister Margaret Thatcher in Britain and President Ronald Reagan in the United States. Neoliberal economic prescriptions at this time called for deregulation, stressing the need for free and open markets as the best way to organize an economy. Neoliberalism advocated liberalization of trade, investment, and financial policies as a way of integrating the global economy, and—it was presumed—of stimulating economic growth. The spread of neoliberal economic thinking in the early 1980s strengthened support for the market-liberal view of the environment, which saw these policies as compatible with environmental protection, as discussed later.

Policies shifted around the world over the course of the 1980s and 1990s to reflect the ascendancy of neoliberal economic thought. While the global economy was widely seen in the 1970s as at least partly responsible for the South's weak economic growth, by the 1980s there was a shift toward the view that global economic integration into free markets was the best path to enhance economic growth. In other words, elites increasingly perceived the domestic failure to open economies as the reason for the economic troubles in developing countries, rather than pointing to a failure of the global economy to provide equal benefits. Institutions like the World Bank and IMF were seen as critical sources of support for the transition to globally integrated free-market economies. The Third World debt crisis was viewed as a crucial test for the neoliberal economic policies of these institutions. In return for emergency loans, the IMF and the World Bank required developing countries to implement economic reforms, commonly known as "structural adjustment." These reforms included liberalization of trade, investment, and exchange-rate policies, devaluation of currencies, government spending cutbacks, and privatization of public enterprises.[37]

The rise of neoliberalism and the realization of the severity of the Third World debt crisis shaped the work of the World Commission on Environment and Development (WCED). The UN General Assembly established this commission in 1984 to examine the relationship between the environment and global economic development. It is commonly known as the Brundtland Commission, after its chair, Norway's prime minister Gro Harlem Brundtland (see box 3.3). Its report, *Our Common Future*, published in 1987, went further than any official international document to provide a new definition of development with the environment at its core. The report, too, was careful to set its analysis in the context of the global political economy.

The Brundtland Report changed the discourse of global environmentalism. The commission tried to chart a middle ground between the North and the South, and between market-liberal and institutionalist views on growth on the one hand, and social green and bioenvironmentalist views on the other. It proposed a global development and environment strategy designed to be palatable to all. It did not see further economic growth and industrialization as necessarily harmful to the envi-

Box 3.3
Gro Harlem Brundtland

 Gro Harlem Brundtland was born in Norway in 1939. Her career is replete with accomplishments. She began as a medical doctor, like her father before her. As a young doctor and mother she won a scholarship to the Harvard School of Public Health, where she earned a Master of Public Health degree. In 1965 she returned to Norway and, while raising a family, worked on children's health issues at the Ministry of Health. But she was drawn to politics, having joined the Norwegian Labour Movement at just 7 years of age. In 1974 she was appointed Minister of the Environment. Seven years later, at 41, she became the youngest and first female prime minister of Norway (as head of the Labour Party). She would go on over a decade-and-a-half period to win three elections and hold onto the prime ministership for about 10 years. Global recognition came as chair of the 1983–1987 World Commission on Environment and Development, where she was instrumental in the creation of the concept of sustainable development as well as the consensus for the 1992 Rio Summit. From 1998 to 2003 she was director-general of the World Health Organization, where she helped to push for a more active and innovative global role for the WHO.

Source: Summarized from the WHO website: www.who.ch.

ronment, and thus did not foresee any necessary "limits" to growth. At the same time, it argued, very much in line with Third World sentiments at Stockholm, that poverty harmed the environment as much as industrialization. This poverty was in large part due to the place of developing economies within the global structure. The best way to move forward, the report contended, was to promote economic growth—not the kind of growth seen in the 1960s and 1970s, however, but environmentally sustainable growth. With this recommendation, the report popularized the term *sustainable development*, which it defined as development that "meets the needs of the present without compromising the ability of future generations to meet their own needs."[38] This definition of development did not pose a fundamental challenge to the neoliberal economic ideology of the time, because it did not suggest a need to slow the pursuit of economic growth or to slow the process of global economic integration.

Other recommendations in the Brundtland Report in some ways reflected the 1970s NIEO agenda. It called for the transfer of environmental technologies and economic assistance so developing countries could afford to pursue sustainable development and economic growth. Other recommendations included calls for states to control population growth, promote education, ensure food security, conserve energy, promote urban sustainability, and develop cleaner industrial technologies. The report challenged some neoliberal economic views—for example, by calling for redistributive measures between rich and poor countries—but the dominance of the neoliberal economic agenda at the time ensured that the global community focused on the fully acceptable compromise of sustainable development.[39] The Brundtland definition of sustainable development is now part of mainstream international and national policies and rhetoric. It has not, though, altered the fundamental dominance of neoclassical economics.[40] In fact, some would argue that the Brundtland Commission's real legacy was to create a concept that secures the hegemony of a coalition of the moderate market-liberal and institutionalist views of environmental management within the global community.

Alongside the work of the Brundtland Commission, the world continued to negotiate global environmental treaties. These were generally in direct response to urgent environmental problems (often generating considerable media attention). These include the 1985 Vienna Convention for the Protection of the Ozone Layer and the 1987 Montreal Protocol on Substances That Deplete the Ozone Layer (see chapter 2), as well the 1989 Basel Convention on the Transboundary Movement of Hazardous Wastes and Their Disposal. The Basel Convention arose in response to the revelations in the mid-1980s of hazardous waste being shipped from rich to poor countries for disposal, raising both ecological and ethical concerns.[41]

Equally important to the diplomatic evolution of environmental discourse in the 1980s were developments in local and national-level environmental movements around the world (many of which began in the 1970s). Three examples are noted here. The West German Green Party, one of whose founding members was Petra Kelly (see box 3.4), was instrumental in putting environmental issues on the German and broader European political agenda in the 1980s. The National Council of Rubber Tappers, a Brazilian NGO, was founded by Chico Mendes (see box 3.5) in 1985 to defend the livelihoods of rubber tappers threatened by deforestation. The Greenbelt Movement in Kenya was established in the 1980s by Wangari Maathai (see box 3.6) as a grassroots women's

Box 3.4
Petra Kelly

 Petra Kelly, born in 1947, was one of the cofounders of the German Green Party in 1979. She became one of the national chairpersons of the German Green Party in the early 1980s, and gained international recognition at this time as a spokesperson for the causes of the Green Party. Kelly was elected to the German parliament in 1983, one of twenty-eight Green MPs elected that year. She authored a number of books and spoke widely on issues like ecology, peace, feminism, and human rights. Kelly received the Right Livelihood Award (known as the alternative Nobel Prize) in 1982. Tragically, she was killed in 1992, the exact circumstances of which remain unclear.

Source: Summarized from the Right Livelihood Award website: www.rightlivelihood.se/recip/kelly.htm.

Box 3.5
Chico Mendes

Chico Mendes (Francisco Alves Mendes Filho) was born in 1944 in the western Brazilian Amazon. Mendes was the son of a poor rubber tapper and himself took up the profession at the age of 9. His efforts to organize rubber tappers to resist nonviolently the destruction of the forests in the Amazon from excessive logging gained international support in the 1970s and 1980s. Mendes founded the National Council of Rubber Tappers in 1985 to further support this cause. The son of a cattle rancher assassinated Mendes in 1988. The subsequent global furor over his death is said to have influenced Brazil's decision to offer to host the 1992 Earth Summit.

Source: Summarized from *Events in the Life of Chico Mendes*, on the Environmental Defense Fund website: http://legacy.environmentaldefense.org/programs/international/chico/chicotimeline.html.

Box 3.6
Wangari Maatthai

Wangari Maatthai, born in Kenya in 1940, was trained in biology and veterinary science and became a professor at the University of Nairobi in the 1970s. Active in the National Council of Women of Kenya from 1976, Maathai went on to chair that organization from 1981 to 1987. As chair she founded the Greenbelt Movement, an organization to promote grassroots action for the environment as a means of improving women's lives. The Greenbelt Movement focused on tree planting and by 1993 had planted over 20 million trees. The movement has been credited with raising environmental awareness in Kenya, gaining global recognition for its achievements. In 1984 Maathai won the Right Livelihood Award for her work with the Greenbelt Movement.

Source: Summarized from the Right Livelihood Award website: www.rightlivelihood.se/recip/maathai.htm.

organization committed to environmental conservation through the planting of trees.

The Earth Summit and the 1990s

In 1989 the UN General Assembly, following the recommendations of the Brundtland Report, passed a resolution to hold another world conference on the environment, the UN Conference on the Environment and Development (UNCED), to be held in 1992 in Rio de Janeiro, Brazil. Significant backdrops to UNCED, popularly known as the Earth Summit, included the ongoing debt crisis in the Third World, the collapse of the Eastern Bloc, and the accelerating pace of economic globalization. Politicians at this time saw green politics as a popular issue, and the axis at the global level for politics of this nature had shifted from East-West to North-South.

The Earth Summit was held on the twentieth anniversary of the Stockholm Conference. The secretary general of the meeting was Maurice Strong, who, as we saw earlier, organized the conference in Stockholm. The Rio Earth Summit was the largest UN conference to date, with 179 countries participating and more than 110 heads of state attending. There were 2,400 nongovernmental representatives as well as another 17,000 in a parallel NGO forum. The recommendations in the Brundtland Report dominated discussions at Rio, especially the notion of sustainable development. Governments at the conference found it politically easy to adhere to the Rio goals of promoting more growth with more environmental protection. Who would deny these goals when presented as mutually compatible?

Many developing countries, however, worried about taking on additional environmental commitments without concrete assurances of economic assistance to fund environmentally sustainable growth. Debates over financial transfers for sustainable development split along North-South lines. Developing countries felt strongly that industrialized countries should foot most of the extra costs of "green growth," and that this should be over and above current levels of assistance. Donors, on the other hand, were reluctant to take on further financial commitments.[42] Related to this theme, developing countries also wanted to ensure receipt of environmental technologies without extra cost, as well as to ensure that industrialized countries would not be able to use environmental regulations to restrict developing-country exports.

What, then, were the main outcomes of the Rio conference? Perhaps most importantly it put environment and development on the agendas of global leaders. It reaffirmed, too, the Brundtland view that more growth will bring a better environment, an idea supported by both the market liberals and the institutionalists. Participating governments adopted and signed the Rio Declaration on Environment and Development, a set of twenty-seven principles outlining the rights and responsibilities of states regarding the promotion of environment and development. These principles included far more of the South's concerns about the right to development than did the Declaration on the Human Environment adopted at Stockholm 20 years earlier. Agenda 21, a 300-page action program to promote sustainable development, was also adopted. The majority of the text of Agenda 21 was negotiated at four separate preparatory committee meetings (PrepComs) in the two and a half years prior to the Earth Summit, and only the last 15 percent was finalized in Rio.[43] The conference also adopted a nonbinding Statement of Forest Principles to promote the sustainable management of tropical, temperate, and boreal forests. It was a compromise document, because it proved impossible to reach agreement on a legally binding forest treaty.[44] The conference also opened two legally binding conventions for signature: the UN Framework Convention on Climate Change and the Convention on Biological Diversity. Further, the conference established the UN Commission on Sustainable Development to monitor and evaluate the progress on meeting the Rio objectives. Negotiations also began at Rio on a treaty on desertification.[45] Finally, Rio was a "trigger" for the restructuring of the Global Environment Facility (GEF), set up to finance efforts in developing countries to protect the global environment (see chapter 7).[46] These global cooperative agreements were very much products of institutionalist influence at the conference.

There were some conspicuous failures at Rio, too. The estimated annual cost to implement Agenda 21 globally was US$625 billion, yet relatively little was pledged at Rio. Moreover, the developed countries were asked to cover just US$125 billion.[47] This amount is roughly equal to 0.7 percent of their total gross national product (GNP), a theoretical target the UN set for development assistance back in the 1970s, and one that most donors have never been close to reaching. In total GEF had only allocated US$4 billion in grants by 2003, although in August 2002, donors pledged another US$3 billion to fund GEF over the next four years.[48]

Many NGO participants at Rio were skeptical of the official agenda. The focus on Brundtland-style solutions, in particular the promotion of economic growth and industrialization as compatible with sustainability, was widely criticized.[49] Social green critics, many of whom were at Rio with more radical NGOs such as Greenpeace and the Third World Network, argued the world community was ignoring the environmental consequences of economic globalization—inequality, industrialization, economic growth, and overconsumption—at its peril. Despite a record number of NGOs in attendance at the global conference, many felt that industry had far more influence on the official agenda, partly because of the close ties to Maurice Strong and industry's funding of the event.[50] A number of critics saw Rio as entrenching a "managerial" approach to solving environmental problems from above, while paying inadequate attention to local solutions from below.[51] These social green critics saw the nod to the role of the poor, women, and indigenous peoples in the official Rio documents as woefully inadequate.

The social green critics integrated some of the bioenvironmentalist concerns with economic growth and overconsumption into their analysis, but did not include a focus on population growth and resource availability. Rather, they stressed the need to reduce global economic inequalities and shift toward more local solutions that genuinely help those most affected by global economic change: women, the poor, and indigenous peoples.[52] It was at Rio that such thought began to fully enter the international diplomatic dialogue on global environmental management.

Despite the critiques and the failure to attract significant extra financing to support the goal of implementing sustainable development, most analysts agree that UNCED did achieve some notable successes, especially in terms of raising environmental awareness among the general public in both the North and South. Five years later a special session of the UN General Assembly, known as the Earth Summit +5, reviewed global progress with the implementation of Agenda 21. The conclusions were largely disappointing: the world had not met most of the Agenda 21 goals. Earth Summit +5 failed to inject new enthusiasm into reaching them.[53]

Johannesburg and Beyond

Economic globalization accelerated in the years following the Rio Earth Summit. Trade and investment as a proportion of the economies of both rich and poor countries grew markedly. The General Agreement on Tariffs and Trade (GATT) concluded the Uruguay Round of trade negotiations in 1994, and the World Trade Organization (WTO) was established in 1995. The WTO has more power to enforce decisions in dispute settlement than the previous procedure under the GATT did, and many viewed this as a prioritization of trade goals above others in the global arena.[54] Economic inequalities have grown since Rio as well, in the view of some a direct result of globalization.[55] A global coalition of nongovernmental forces emerged in the late 1990s to oppose globalization. These groups began to protest at global economic meetings, such as the annual meetings of the IMF, World Bank, and WTO. The largest and most significant was at the Seattle ministerial meeting of the WTO in late 1999. Antiglobalization protesters rallied against the environmental and social fallout from globalization, blaming global economic institutions for aggravating inequalities.[56] The WTO failed to begin a new round of trade negotiations at Seattle. The WTO finally launched the Doha Round of trade negotiations in the fall of 2001, at WTO meetings held in Doha, Qatar.

The antiglobalization sentiment of activists, as well as the steadfast support for globalization among governments and international economic institutions, were continually in the background during the preparations for the 2002 World Summit on Sustainable Development (WSSD), held in Johannesburg, South Africa, 10 years after the Rio Earth Summit (commonly called Rio +10). The goal was to evaluate the progress toward the Rio goals and set concrete targets to improve implementation. It also aimed to build on the International Conference on Financing and Development, held in March 2002 in Monterrey, Mexico, as well as to develop a strategy to implement the UN's Millennium Development Goals.[57] The WSSD was the biggest UN meeting ever. Over 180 nations and 100 heads of state attended. There were over 10,000 delegates, 8,000 civil-society representatives (i.e., representing nongovern-mental interests), 4,000 members of the press, and countless numbers of ordinary citizens. The secretary general was Nitin Desai, an Indian diplo-mat with the UN who served as deputy secretary general of UNCED under Maurice Strong, and who continued to work at the UN at the undersecretary general level on policy coordination and sustainable development.[58]

Much like Rio and Stockholm, the conference adopted official docu-ments about the need for global sustainability as well as a plan for imple-mentation. These are the Johannesburg Declaration on Sustainable Development, which outlines both the challenges and general commit-ments for the global community, and the Johannesburg Plan of Imple-mentation, an action plan for implementing these goals. Negotiations for these documents began long before the meeting, yet one-quarter of the text was still in dispute at the start of the conference. One of the most contentious issues, as was the case in Rio, was financing. But new and equally difficult items were added at Johannesburg: globalization's role in sustainable development, as well as timetables and specific targets for meeting goals.[59] In the end, the Johannesburg Declaration on Sustainable Development was somewhat different from previous global declarations on environment and development. Economic inequality was recognized as a major problem not only in terms of justice and human well-being, but also in terms of global security. And globalization was mentioned as a new challenge for achieving sustainable development.[60]

The Plan of Action, a sixty-five-page document, was much less ambi-tious than Rio's Agenda 21 and largely restated the areas where more action was needed to meet the Rio goals, as well as other previously agreed-on goals such as the Millennium Development Goals and devel-opment financing goals.[61] The Plan of Action did include a few more spe-cific goals and targets, focusing on several main priorities, such as water, agriculture, energy, health, and biodiversity. The Commission on Sus-tainable Development, created after Rio, was reaffirmed as the agency that would monitor progress on this implementation.

The conference also highlighted and promoted a new instrument for implementing sustainable development. These are public-private agree-ments (dubbed "Type II"—"Type I" being government to government) between governments, NGOs, and business. These are voluntary agreements that involve stakeholders directly. The aim is to encourage the transfer of funds and technology to areas of critical need. Several hundred such partnerships were identified prior to and during the Summit.[62] Greenpeace, for example, announced a partnership with the World Business Council for Sustainable Development to promote action to mitigate climate change. This focus on public-private partnerships gave the Johannesburg Summit a greater focus on the role of business in promoting sustainable development than either Stockholm or Rio had. The Plan of Action mentions the need to promote corporate accountability and responsibility, and the Johannesburg Declaration stresses the duty of business to promote sustainable development.

Was Johannesburg a success? Many business participants saw the outcome as constructive and realistic. Many nongovernmental participants were disappointed with the "weak" targets and timetables, noting that the conference added little to global efforts to enhance sustainability.[63] New targets were set on access to clean water and sanitation, as well as on halting depletion of fish stocks, but beyond these, not much new was added in terms of global commitments. The debate over the role of corporate globalization was divided, with industry and most governments on the one hand seeing it as positive, and many environmental and development NGOs on the other seeing it as negative. In the end the official documents make bland and noncommittal statements about the importance of this issue without offering guidance on how to proceed.

Some argue that the Johannesburg Summit highlights the waning impact of environmentalism in global discourse.[64] Part of the reason that many of the debates at the WSSD did not make much progress was the clash of diverse views on the causes and consequences of global environmental change. Social greens and bioenvironmentalists were able to present their cases much more effectively than in previous global environmental meetings. This raised tough issues at the negotiating table, such as globalization, corporate accountability, exhaustible resources, and consumption. Yet the overall dominance of the views of market liberals and institutionalists in the official proceedings ensured that such "talk" did not result in concrete targets or plans of action. The UN, partly because Johannesburg disappointed so many, is planning to focus more on implementation and yearly reviews of progress, rather than on holding another large environmental summit anytime soon.

Global Environmental Governance

The global discourse on environment and development shapes, and is shaped by, institutions and policies—from global to local, and public to private. The history of global environmental discourse and diplomacy, as presented in the first part of this chapter, highlights the fact that over time there have been significant changes to the key actors. In today's world, not just states, but also international institutions, firms, markets, NGOs, communities, and individuals shape global environmental governance.[65] These actors also employ an increasingly diverse range of policy tools. Thus, it is hardly surprising that we see a wide range of views on how to best pursue global environmental protection. Here, to provide a more comprehensive view of the globalization of environmentalism, we provide a brief discussion of states, regimes, NGOs, and corporations in the context of emerging patterns of global environmental governance. We add a deeper understanding of the impact of these actors on global environmental management throughout the rest of the book.

States

States are at the core of global environmental governance. This is natural given that sovereignty, consisting of both rights and duties, is the guiding principle of international politics. The Treaty of Westphalia, signed in 1648, gave a state supreme authority to act within its territory. Foreign powers could not dictate what states can and cannot do within their borders. At the same time, to preserve these sovereign rights, states are obliged to respect the sovereignty of other states as well as protect their own citizens. States, as a glance at the wars of the last 350 years shows, do not always respect the sovereignty of other states. Nor do states always protect their citizens. Nevertheless, all states have guarded their sovereign rights with vigor and resolve, a fundamental feature of global environmental politics.[66]

There are obvious tensions between the sovereign-state system and the natural environment. Perhaps the most visible is that environmental problems commonly spill across the political boundaries of states. The actions needed to manage the global environment—to act in the interests of the planet—may well clash with state interests. This partly explains the global community's struggle to deal with problems like climate change. But if the sovereign-state system grants states the right to control activities within their borders, what are their duties toward other states if their activities damage the environments of other states or the global commons? Principle 21 of the Stockholm Declaration addresses this question, declaring that states do have a right to exploit their own resources or pollute their own environments, but only so long as this does not affect other states directly.[67] It is, however, difficult to prove that the actions of one state harm the environment of another state (or states) or the global commons. Can it be proven, for example, that deforestation in Costa Rica harms the United States because it decreases the global capacity to sink carbon dioxide? And who is to blame: the Costa Rican government, Costa Rican cattle ranchers, or the consumers of steaks and hamburgers?[68] Because of this difficulty of proof, in practice little beyond moral pressure tends to be applied to hold states to this principle.

A second tension between the sovereign-state system and the natural environment is the tendency for environmental problems to develop over a long period of time. The existence of some problems, such as depletion of the ozone layer by CFCs, took years not just to discover, but also to prove with enough certainty to convince states to act. Environmental problems also tend to be circular and dynamic in nature. They rarely have a single cause, and efforts to deal with one part of an environmental problem can easily create new problems in the future. This long-term and circular nature of environmental problems contrasts sharply with the short-term and results-oriented political lives of those in power in democratic nations. It is naturally difficult for politicians to gain the funding and support for long-term action with political cycles typically lasting 5 years or less.[69]

Many recognize the contradictions and tensions of the state system for effective global environmental management. Most assume it is unrealistic, perhaps even counterproductive, to try to replace it (with, for example, a federal world government). Instead, virtually every state has now created an environment department with national and global responsibilities. States have embraced economic globalization, which strengthens the power of the global market to transfer environmental funds and technology across state borders (although, as we have seen, many argue that this also transfers more environmental harms across borders). States have passed the authority for some environmental issues to global institutions. And finally, states have relinquished some sovereign rights by signing onto hundreds of global environmental agreements. States still dominate global environmental governance. But, as the next section shows, international institutions and regimes have grown in importance.

International Institutions/Regimes

Since Stockholm, global institutions have become increasingly responsible for global environmental protection. The UNEP is a catalyst for global action among states as well as other actors to mitigate specific environmental problems.[70] It does this by supporting research on and negotiation of new international environmental agreements.[71] Other arms of the UN also promote sustainable development. The Commission on Sustainable Development monitors the implementation of Agenda 21 as well as other internationally declared environmental goals and targets.[72] The UN Development Programme (UNDP), the World Health Organization (WHO), the World Bank, and the Food and Agriculture Organization (FAO), just to mention a few, have also—if in a less direct fashion—shaped global environmental management. Table 3.1 outlines some of the key international organizations that influence global environment and development governance. In addition, many of the Regional Commissions of the UN have devised and implemented environmental strategies. The Organization for Economic Cooperation and Development (OECD) has adopted and applied legally binding regulations on the environment. A few of the many intergovernmental groups that consider environmental policies and strategies include the Group of 8, the African Ministerial Conference on the Environment, the Council of Arab Ministers Responsible for the Environment, and the European Union Environment Ministers.[73]

International environmental regimes encompass established rules of international cooperation, either through a formal international treaty or agreement, or through a less formal set of rules and norms (established practices that states generally adhere to).[74] States have signed and ratified a remarkable number of global environmental agreements between the Stockholm and Rio conferences (see table 3.2). In 1972 there were only a couple of dozen multilateral environmental treaties, whereas today there are several hundred.[75]

Are these agreements and regimes working? Chapter 2 showed that some, such as the ones to solve depletion of the ozone layer, have been reasonably successful. At a minimum the secretariats to the conventions are able to disseminate critical information as well as influence a state's level of concern about an environmental issue. Through aid and training the secretariats also help to raise the national capacity of developing countries struggling to comply with global agreements. Many observers, however, remain skeptical of the value of international agreements—arguing that many do not in fact alter state and corporate behavior.[76] States that do not sign or ratify are not bound by a treaty, creating obvious problems of free riders (a general drawback of international law). It is exceedingly difficult to design effective agreements and enforce compliance.[77] It is also becoming increasingly tricky to coordinate the plethora of sometimes overlapping institutional arrangements for global environmental management.[78] It is, moreover, often unclear that full compliance with a treaty will automatically improve environmental conditions (a point commonly made about the Kyoto Protocol).

Nongovernmental Organizations and Activist Groups

We define NGOs as organizations that are autonomous from governments, and are generally nonprofit (thus excluding corporations), though the lines are not always clear. International NGOs can include a wide variety of types of groups, such as large transnational organizations (like Greenpeace or Friends of the Earth International), regional groups (like the Third World Network), and local and grassroots groups and international networking groups that facilitate coordination of such groups (like the International Rivers Network and the many local NGOs around the world networked with it), as well as research institutes (like the International Institute for Sustainable Development, or the Worldwatch Institute). Some of these groups have significant budgets. The budget of Greenpeace (including national offices) was close to US$150 million in

Table 3.1
Intergovernmental organizations: Voting rules and revenue sources

Organization	Voting rules	Revenue source
United Nations	Each member of the General Assembly has one vote; decisions are made by a two-thirds majority of the members present and voting.[1]	Member contributions, as apportioned by the General Assembly.[2]
United Nations Environment Programme	Governing Council is composed of 58 members of the UN elected by the UN General Assembly for 3-year terms, taking into account the principle of equitable regional representation. Decisions are adopted by majority of members present and voting.[3]	The costs of servicing the Governing Council and providing the small secretariat are borne by the regular budget of the United Nations. The Environment Fund provides financing for environmental programs; altogether 154 countries have made at least one voluntary contribution.[4]
United Nations Development Programme	The Executive Board of UNDP, consisting of 36 members, is elected by the Economic and Social Council for 3-year terms. Since 1994, decisions have always been adopted by consensus.[5]	Donor support, principally from members of the OECD DAC, plus local resources provided by host governments and reimbursement for services provided by the program.[6]
World Bank	Voting power for members includes "membership votes" (identical for all members) plus additional votes based on shares of Bank stock held. Voting power equals a member's votes expressed as a percentage of the total number of votes held by all shareholders.[7]	20% of member subscriptions ("subscribed shares") of authorized capital stock in the Bank; guarantee commissions and interest on loans made by the Bank; buying and selling of securities in which the Bank has invested (including borrowing the currency of any member).[8]
Global Environment Facility	Council consists of 32 member "constituencies"; 16 are represented by developing country members (6 each from Africa and Asia, 4 from Latin America), 14 by developed countries, and 2 by nations with transitional economies. Decisions are made by consensus; when this is not possible, voting is done by "double-majority" (a majority of participating countries plus 60% donor support).[9]	Funds from 32 donor nations are committed every 4 years in the "GEF Replenishment" process.[10]

International Monetary Fund	Voting power determined by quota, based on relative size in the world economy.[11]	Quota subscriptions from each member country, assigned based on its relative size in the world economy.[12]
World Trade Organization	Follows GATT (1947) consensus model: decisions are adopted if no member present formally objects. When voting is required, each member has one vote.[13] Common practice is the "Green Room meeting," where the chairperson of a negotiating group consults privately with delegations, either individually or in groups, in an attempt to forge a compromise.[14]	Member contributions, established according to a formula based on their share of international trade; management of trust funds contributed by members; rental fees and sales of WTO print and electronic publications.[15]

1. & 2. Charter of the United Nations, Art. 18, Sec. 1 & 2, and Art. 18. Available at www.un.org/aboutun/charter.
3. Rules of the Governing Council. Available at www.unep.org/Documents/Default.asp?DocumentID=77&ArticleID=1157.
4. Financing of UNEP: Regular Budget. Available at www.unep.org/rmu/html/fund_regular.htm. And Financing of UNEP: Environment Fund. Available at www.unep.org/rmu/html/fund_environment.htm.
5. Executive Board of UNDP and UNFPA. Available at www.undp.org/execbrd/pdf/eb-overview.PDF.
6. UNDP budget estimates for the biennium 2004–2005. Available at www.undp.org/execbrd/pdf/dp03–28e.pdf.
7. World Bank Board of Directors: Voting Powers. Available at http://web.worldbank.org.
8. IBRD Art. IV, Sections 4 and 5 and Art. II, Section 5. Available at http://web.worldbank.org.
9. GEF, *Handbook of International Cooperation on Environment and Development*. Available at www.greenyearbook.org/igo/gef.htm.
10. GEF, *Replenishment*. Available at www.gefweb.org/Replenishment/replenishment.html.
11. & 12. IMF Quotas: A Factsheet. Available at www.imf.org/external/np/exr/facts/quotas.htm.
13. WTO: Module 1 FAQs. Available at www.wto.org/english/thewto_e/whatis_e/eol/e/wto01/wto1_51.htm#note3.
14. Understanding the WTO: The Organization. Available at www.wto.org/english/thewto_e/whatis_e/tif_e/org1_e.htm.
15. WTO Secretariat Budget for 2003. Available at www.wto.org/english/thewto_e/secre_e/budget03_e.htm.

Table 3.2
Chronology of international environmental cooperation (summary of major initiatives)

1954	International Convention for the Prevention of Pollution of the Sea by Oil is adopted. 71 parties; entered into force in 1958.
1972	United Nations Conference on the Human Environment is convened in Stockholm. Stockholm Declaration is adopted.
1972	Convention on the Prevention of Marine Pollution by Dumping of Waste and Other Matter (London Dumping Convention) is adopted. 80 parties; entered into force in 1975.
1973	International Convention for the Prevention of Pollution from Ships (MARPOL) is adopted. 125 parties; entered into force in 1983.
1973	Convention on International Trade in Endangered Species (CITES) is adopted. 162 parties; entered into force in 1975.
1973	Creation of the United Nations Environment Programme (UNEP).
1989	The Basel Convention on the Control of Transboundary Movements of Hazardous Wastes and Their Disposal is adopted. 158 parties; entered into force in 1992.
1991	Establishment of the Global Environment Facility (GEF). 176 member countries.
1992	United Nations Conference on Environment and Development is convened in Rio de Janeiro. Agenda 21 and Rio Declaration are adopted.
1992	United Nations Framework Convention on Climate Change is adopted. 188 parties; entered into force in 1994.
1992	United Nations Convention on Biological Diversity is adopted. 187 parties; entered into force in 1993.
1992	Creation of the United Nations Commission on Sustainable Development (CSD). 53 members.

1979   Convention on Long Range Transboundary Air Pollution is adopted. 49 parties; entered into force in 1983.

1979   Convention on the Conservation of Migratory Species of Wild Animals is adopted. 84 parties; entered into force in 1983.

1982   United Nations Convention on the Law of the Sea is adopted. 142 parties; entered into force in 1994.

1985   Vienna Convention for the Protection of the Ozone Layer is adopted. 185 parties; entered into force in 1988.

1987   Montreal Protocol on Substances That Deplete the Ozone Layer is adopted. 184 parties; entered into force in 1989.

1987   The Brundtland Report is published.

1994   International Convention to Combat Desertification is adopted. 187 parties; entered into force in 1996.

1997   Kyoto Protocol on Climate Change is adopted. 120 parties; not yet in force.

1998   The Rotterdam Convention on the Prior Informed Consent procedure for Certain Hazardous Chemicals and Pesticides in International Trade is adopted. 56 parties; entered into force in 2004.

2000   Cartagena Protocol on Biosafety is adopted. 51 parties; entered into force in 2003.

2001   Stockholm Convention for the Elimination of the Persistent Organic Pollutants (POPs) is adopted. 43 parties; entered into force in 2004.

2002   World Summit on Sustainable Development held in Johannesburg. Johannesburg Declaration and Plan of Implementation are adopted.

2001. The budget of the WWF (including national organizations) in 2000–2001 was around US$350 million. This compares favorably to the UNEP's budget in 2001 of about US$120 million.[79]

There has been, as with international environmental agreements, a significant increase in the number of international NGOs over the past few decades. In 1990 there were 6,126 NGOs with some international characteristics. This jumped to 23,135 in 1996 and to 43,958 by the year 2000.[80] To some extent this explosion of participation of civil society—on issues ranging from protection of the environment, to the rights of minorities and indigenous peoples, to the fair treatment of women—is evidence of a growing skepticism toward the efficacy of states in the current global political economy.[81] Environmental NGOs increasingly participate in the negotiation processes for international environmental treaties. Some have gained UN accreditation through the Economic and Social Council (today about 1,000), but others are let in as observers (just about anyone who asks can attend). This participation of NGOs in the UN treaty process goes back to the formation of the UN—NGOs were recognized in 1948 as important actors that states could officially consult. NGOs cannot vote in formal UN settings or in environmental treaty negotiations/Conferences of the Parties. But sometimes they are included on national delegations. And sometimes they are able to address the plenary sessions as well as most of the smaller negotiating meetings and workshops where state officials generally hammer out details.

International environmental NGOs can have particular advantages that help them influence global agreements.[82] They can develop specialized knowledge and expertise on specific issues, which they can then use to challenge mainstream scientific assumptions. Many state delegates do not have particular expertise, and in such cases NGO interpretations can be very important in shaping states' positions. Accreditation also allows some international environmental NGOs to gain direct access to global decision-making processes.[83] In addition, these NGOs sometimes form alliances with governments (especially those of developing countries), and help them with scientific data and negotiations. For example, in the hazardous-waste-trade negotiations, Greenpeace formed a strong alliance with G-77 countries to push for a waste-trade ban.[84] In fulfilling these roles, environmental NGOs enhance the transparency of dominant actors—that is, the states, international organizations, and corporations—although it is important to point out that NGOs can also suffer from bias and opaque procedures. In negotiations wealthier Northern NGOs may also dominate NGOs from the South. Other actors can co-opt NGOs at times, and their presence at negotiations may simply "legitimate" a state-based treaty process.

Another important role of environmental NGOs is to educate and raise global environmental consciousness, even transforming global culture.[85] Daring stunts can capture global media attention and help make complex scientific issues accessible to ordinary people. Greenpeace was able to change the prevailing image of whaling from that of a heroic battle to a perception of the slaughter of the innocent.[86] NGOs can influence other nonstate actors, such as corporations. They can empower and help alleviate poverty in local communities in the South. And they can help translate local concerns into the global language of treaties and financial assistance.[87]

Nongovernmental participation in shaping global environmental politics also encompasses efforts by groups and movements that may not focus exclusively on the environment, but that do have an interest in particular environmental issues in their broader agenda. These include many indigenous peoples' organizations, which actively participate in global negotiations and dialogue on environmental issues that affect them. The Inuit Circumpolar Conference, for example, was instrumental in efforts to bring about a global treaty on POPs (the Stockholm Convention on Persistent Organic Pollutants).[88]

The emergence of Green parties has also influenced environmental governance nationally as well as globally. The West German Green Party, as mentioned earlier, was a national political party that aimed to further goals of a number of social movements, including obviously the environment, but also civil rights, peace, and women's rights. The success of this party in the 1980s inspired the development of national Green parties first in other European countries and then throughout the world. Though they are national political parties, they are linked through an international network with a Global Green Charter.[89]

Corporations

Firms have also increased their role and influence in the global environmental arena in recent years, in part as a response to the enhanced role of environmental NGOs. Both business advocacy groups, such as the International Chamber of Commerce (ICC) and the World Business Council for Sustainable Development (WBCSD), and individual transnational corporations (TNCs), like DuPont and Monsanto, have been actively engaged in global discussions on the environment and development.

Like environmental NGOs, corporate actors also play a diplomatic role in an effort to enhance interstate cooperation on environmental issues. Lobbying before state delegations head to international environmental negotiations has enabled firms to influence governmental positions from behind the scenes.[90] Business actors have also increasingly begun to lobby at the international level as well, through industry associations and industry representatives with observer status at global environmental meetings. The presence of these actors at global negotiations is now routine—for example, during negotiations on the waste trade, climate change, ozone depletion, biosafety, and persistent organic pollutants. Business actors make fewer public interventions than environmental NGOs do (for example, in plenary sessions and smaller meetings), but they are active in the corridors, lobbying and shaping the positions of states. Business representatives also end up on national delegations. This trend can be seen in both positive and negative lights. Market liberals and institutionalists argue that business groups should take part in negotiating treaties that will affect them directly, because this will ensure greater compliance later. Critics, in particular social greens, argue that the increasing influence of corporate environmental lobbyists means international environmental law favors economic goals over environmental ones.

Corporate actors also influence global environmental governance through other international forums as well. Firms have helped to shape private forms of global environmental governance, such as industry-set voluntary codes of environmental conduct at both the national and international levels.[91] TNCs also exert influence in more diffuse ways, such as through their role in the global discourse over defining key terms, such as *sustainable development*,[92] and through their economic weight, which influences state decisions in an indirect way.[93] We expand on these key themes in more detail in chapter 6.

Conclusion

The globalization of environmentalism should be seen as a complex and fluid process with no neat beginning, middle, or end point. Undeniably important were groundbreaking publications like *Silent Spring*, *The Population Bomb*, *The Limits to Growth*, *Small Is Beautiful*, the Founex Report, and the Brundtland Report. So were the watershed global conferences at Stockholm, Rio, and Johannesburg. So were the hundreds of international environmental agreements on issues like ozone depletion, hazardous waste, climate change, and desertification. And so were the emergence of local and global activist groups around the world. These developments all took place in a broader political and economic context. To fully understand the rise of sustainable development and the current structures and norms (i.e., established practices and understandings) of environmental governance, we have argued that it is equally critical to examine the broader trends in the global political economy. Global economic shifts—in the generation and distribution of wealth—have been key in influencing the evolution of perspectives not just on the economy but also on the relationship between the global economy and the environment.

Globalization and the global interaction of states, institutions, firms, markets, communities, and individuals will continue to shape and reshape global environmental governance for decades to come. As in the past, the arguments of market liberals, institutionalists, bioenvironmentalists, and social greens will continue to swirl around each other. The result of the globalization of environmentalism has to some extent split the current international debate over how to best manage the global environment into two broad coalitions: on the one side, market liberals and institutionalists calling for more effective implementation of sustainable development; on the other side, bioenvironmentalists and social greens calling for a new global political economy built on the principle of ecological and social sustainability. For now, however, the dominance of the "Brundtland compromise" of "sustainable development" gives institutionalists and market liberals the upper hand in the global community. Many NGO and activist groups that draw more from social green and bioenvironmentalist thought will no doubt continue to be highly critical of this compromise, especially the assumption that economic growth can (and should) continue indefinitely. It is important, however, to avoid oversimplifying the cases for and against sustainable development, because there are distinct differences in assumptions, arguments, and solutions across the four worldviews. The next chapter explores in detail the different assumptions and arguments about the global ecological consequences of economic growth.

Economic Growth in a World of Wealth and Poverty

Does wealth lead to better or worse environmental conditions? Does poverty lead to ecological neglect as people struggle to survive, or do the poor place less of a burden on the environment than the rich? There are no easy answers to these questions. They are intensely debated, and the conclusions one reaches will depend very much on one's worldview. The question of economic growth is at the core of most of the debates over the global economy and the environment, because it is intricately tied to debates over wealth and poverty and their relationship to the quality of the natural environment. This chapter examines these debates in detail, thus providing a foundation for understanding the debates about the other global economic issues we discuss later in this book, such as trade, investment, and aid.

Debates over growth at first glance appear to be polarized into two camps, much like the debate on globalization: those who see growth as a positive force for the environment, primarily market liberals and institutionalists; and those who view growth as largely negative, primarily bioenvironmentalists and social greens. But when we look carefully at how each group examines the questions of poverty, wealth, and population in relation to one another, and in relation to economic growth, we see that the debate is more complex than a simple two-camp split.

Wealth and Poverty for Market Liberals and Institutionalists

The rate of economic growth is a core measure of a country's economic performance. Typically it is measured by the rate of growth of the gross national product (GNP) or gross domestic product (GDP) (see box 4.1), which, with UN support, became standard measures for "development" in national income accounting after World War II. A rise in GNP or GDP implies growth of industrial production and in turn growth of consumption. This in effect means that more consumption equals development, as more needs and wants are met, which presumably betters the lives of ordinary people. The loose correlation between high economic growth and indicators such as life expectancy, health, and literacy adds statistical weight to this view. Global economic institutions such as the World Bank and International Monetary Fund, both of which tend to hold market-liberal outlooks on environmental issues, rely heavily on GNP and GDP figures to rank countries' economic performance.

Assessed with these measures, the global economy has performed well over the past 50 years. Global GDP has jumped from US$7.8 trillion in 1960 to US$35 trillion in 2002 (in constant 1995 dollars; see figure 4.1). Global GDP per capita has grown from approximately US$2,600 to US$5,600 in that same period (see figure 4.2). These global figures, however, hide some noteworthy variations across regions. East Asia's GDP grew from US$135 billion in 1960 to US$1.9 trillion in 2002 (with average annual growth rates well over 5 percent), while Latin America's GDP grew from US$430 billion to US$2 trillion over that same period. Meanwhile sub-Saharan Africa only went from US$106 billion to US$395 billion, and South Asia from US$105 billion to US$681 billion (in constant 1995 U.S. dollars).[1]

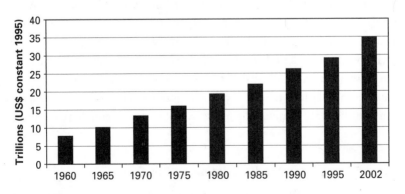

Figure 4.1
Global GDP. *Data source*: World Bank World Development Indicators: www.worldbank.org/data.

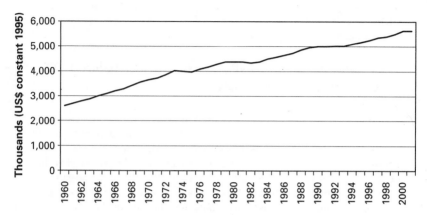

Figure 4.2
Growth in global GDP per capita. *Data source*: World Bank World Development Indicators: www.worldbank.org/data.

Some institutionalists have tried to push the global community to recognize the limitations of relying on total global GDP and global GDP per capita as the sole measures of development. Since 1990 the UN Development Programme's Human Development Index (HDI) has challenged the simple GDP measures, trying to create a more holistic assessment of economic performance and development. The 2001 Human Development report explains the idea of human development:

Human development is about much more than the rise or fall of national incomes. It is about creating an environment in which people can develop their full potential and lead productive, creative lives in accord with their needs and interests. People are the real wealth of nations. Development is thus about expanding the choices people have to lead lives that they value. And it is thus about much more than economic growth, which is only a means—if a very important one—of enlarging people's choices.[2]

The HDI ranks countries based on their performance on a range of indicators: life expectancy at birth, educational attainment (including literacy rate), and GDP per capita (see box 4.1). The HDI rankings are different from simple GDP per capita rankings (see figures 4.3 and 4.4), though the differences are not dramatic. The media reports these rankings widely and countries like Canada and Australia have prided themselves on consistently high rankings. The HDI rankings have helped to expand the mainstream analysis of economic performance somewhat, and over 120 countries now publish national human development

Box 4.1
Measures of economic development

Gross National Product (GNP), Gross Domestic Product (GDP), and Gross National Income (GNI)

GNP and GDP are the most widely used indicators of economic growth. GNP measures the total monetary value of all goods and services produced by a country's resources, regardless of where production takes place. GDP measures the total monetary value of goods and services produced within a country's territory. So what is the difference between these measures?

To be more precise, GNP measures the value of goods and services produced by a country's citizens or nationals, whether or not they live in that country. The example of the accounting of a Japanese auto plant in the United States will help to illustrate the differences between GDP and GNP. Presuming most of the workers are American, U.S. GNP only includes the part of the value of production that accrued to American workers—that is, the employees' wages are part of GNP but the profits are not (they are counted in Japan's GNP). For GDP a transnational corporation's income is counted in the country where the income was generated. In the above example all of the value of the plant's production is included in the U.S. GDP and none is included in the Japanese GDP.

Both indicators give a rough idea of a country's economic activity. Nevertheless, over the past decade or so most countries have switched from GNP to GDP, partly because the UN has tried to standardize economic accounting across countries (the UN's System of National Accounts uses GDP). GDP per capita is now the most common measure of "development" or "progress," since this captures the amount of wealth available per person in a country.[1] The switch to GDP as a measure of national income did result in some major shifts in that TNC income is now wholly counted in the country in which the production takes place, raising the GDP of some countries that are host to a large number of TNCs.

Some organizations, such as the World Bank, have replaced the use of the term GNP with GNI. GNI is the income of the sectors that produced the goods and services measured in GNP, and so by definition it is identical to GNP in value, though technically it is a measure of income rather than production.

Human Development Index (HDI)

HDI is a measure of development that ranks countries according to their performance on indicators linked to longevity, knowledge, and income. Life expectancy at birth is the indicator used for measuring longevity. Educational attainment is the knowledge indicator, and includes a two-thirds

Box 4.1
(continued)

weight to the adult literacy ratio and a one-third weight to primary, secondary, and tertiary enrollment ratios. Income is measured in real GDP per capita. This is determined by counting income up to US$5,000 at full value, and after $5,000 its value diminishes in the index. In this way the GDP per capita is adjusted for its real value (i.e., what it can purchase) rather than being assessed at official exchange rates. HDI then measures the human condition in a broader sense than does income alone. It is widely accepted among development thinkers and practitioners as a valid measure of development, even though GNP/GDP measures are still predominant.

1. We are grateful to Torben Drewes for his comments on this section.

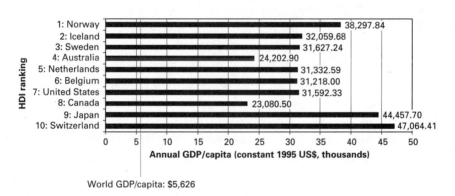

World GDP/capita: $5,626

Figure 4.3
HDI and GDP, 2001, top ten HDI-ranked nations. *Data source*: UNDP, 2003; World Bank World Development Indicators: www.worldbank.org/data.

reports. Nevertheless, GNP and GDP figures still tend to dominate national and global economic policymaking.

There are, then, some mild differences among market liberals and institutionalists on how best to measure development. But both draw on neoclassical economics to address questions of the implications of economic growth. For neoclassical economists, growth within an economy occurs through the functioning of the market. The economy is seen as a circular flow system. That is, there is a closed loop between firms and households in the wider macroeconomy. Firms produce goods and services,

Figure 4.4
HDI and GDP, 2001, bottom ten HDI-ranked nations. *Data source*: UNDP, 2003; World Bank World Development Indicators: www.worldbank.org/data.

and households (consisting of consumers) consume them. These same households in turn make the means of production (labor in the form of workers and capital in the form of investment) available to the firms again. The firms continue the flow through the circle by producing more goods and services, and the flow moves forward indefinitely. The macro-economy is thus seen as self-sustaining. Economic growth occurs as more goods and services are produced. In this way the overall economy is thought to be capable of growing indefinitely (see figure 4.5).

How does the natural environment fit into this circular flow system that generates growth? The macroeconomy is seen as an all-encompassing system, within which everything else is a part, including the ecosystem. This model does not factor in the natural environment as part of the circular flow between firms and households. Rather it assumes the natural environment is static and infinitely available. Resources are thus not a crucial factor of production, because growth is seen to occur primarily as a result of inputs of capital and labor.

Neoclassical economics primarily views environmental problems as arising from a type of market failure known as negative externalities. Market failure broadly refers to specific instances where markets on their own do not allocate resources efficiently. A negative externality is a sit-uation where the market does not account for the impact of an economic activity on those not directly engaged in that activity. Pollution is widely regarded as a negative externality. Though there may be cleaner pro-duction processes that do not pollute, often firms choose the less clean

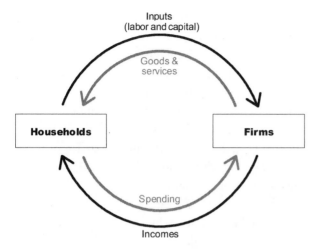

Figure 4.5
The economy as a circular flow system. *Source*: Adapted from Dornbusch, Fischer, and Sparks 1993, 28.

methods because they are cheaper to adopt. This situation then represents a failure of the market to operate efficiently, since all the costs of production (i.e., the cost of production plus the costs of pollution) are not fully accounted for. In this case, the health and cleanup costs of the pollution are "externalized" because they are not borne by the firm, but by governments and the general public.

Often negative externalities occur because environmental costs do not receive a sufficient monetary value. In a free-market system most neoclassical economists see negative externalities as relatively rare and in cases where they do occur, they are assumed to be reversible with proper regulation, taxes, or other economic tools. This view of the natural environment in neoclassical economics is not surprising, given that the field emerged in the nineteenth century, a time of frontier economies, where natural resources and sinks for pollution seemed limitless.

Natural Resources and Sinks Need Not Be a Constraint
Today market liberals and institutionalists still largely accept the foundational principles of neoclassical economics. Yet, over the past 30 years, many have modified the parameters, seeing pollution as a serious problem, and accepting that certain natural resources are finite. This

challenges the neoclassical assumption that environmental harm is a rare market failure. Economic policies must therefore account for externalities to ensure effective environmental management, particularly for problems like air pollution and deforestation. These thinkers note, for example, that firms commonly pollute the atmosphere without penalty or full cost (individuals, governments, and the global commons often absorb the cost). They point out, too, that firms regularly obtain logging licenses with only a token charge for the environmental or social value of the original forest ecosystem (which explains why tropical wood is so inexpensive, despite the threat deforestation poses to much of the globe's biodiversity).

The subfield of environmental economics, which emerged in the 1970s, specifically seeks to address market failures and negative externalities related to natural resources and pollution. It works to price natural resources and sinks more accurately to avoid market failures, and to bring them more effectively into the circular flow system.[3] Environmental economists seek to correct environmental externalities through proper tax levels as well as other economic incentives, such as user fees. Taxes, for example, could be levied against polluters as a disincentive to pollute. User fees, such as charges for garbage collection or for access to protected areas, can provide the state with revenues with which it can address environmental problems. Environmental economics assumes that such market failures are a key source of unsustainable development, yet importantly these failures are correctable. While this approach has contributed to more elaborate models to account for environmental change and scarcity, many of the original tenets of neoclassical economics remain.

There is still a widespread assumption, for example, that as resources and sinks become genuinely scarce, prices for them will rise, which in turn will foster the ingenuity to find alternatives. This could involve new resources to replace the scarce ones, producing goods with less resources or waste, or discovering new ways to manage pollution, including the use of taxes and other market mechanisms. This view still places great faith in the ability of humans to use existing resources more efficiently (by improving technology), to manage existing resources effectively through environmental policy, or to discover "new" resources to drive economic growth. Economist Julian Simon, challenging the bioenvironmentalist view of resource scarcity, argued that "the supply of natural resources is not finite in any economic sense, which is one reason why their cost can continue to fall indefinitely." For him, with current and future knowledge "we and our descendants can manipulate the elements in such fashion that we can have all the raw materials that we desire at prices ever smaller relative to other goods and to our total incomes."[4]

The depletion of natural resources and the filling of waste sinks, then, may pose some environmental and economic problems for market liberals and institutionalists. Yet from their standpoint, with the aid of ingenuity and appropriate policies and institutions, these should not be acute constraints on economic growth and the accumulation of wealth. Market liberals place faith in individuals and firms to come up with solutions to scarcity. Institutionalists stress the need to foster such ingenuity through government and institutional efforts. The implication is the same for both, however: there is no need whatsoever to abandon the growth-based model of development. Growth can and must go on indefinitely for the sake of humanity as well as for the sake of the global environment. Growth just needs to be made more efficient and less polluting. Pricing resources and sinks properly can ensure that this occurs.

Growth Brings a Cleaner Environment: The Environmental Kuznets Curve

How, then, does steady economic growth ensure a cleaner future? The answer, say market liberals and institutionalists, is best illustrated with the environmental Kuznets curve (or EKC, also known as the inverted-U curve).[5] This EKC shows that economic growth will sometimes harm the environment in the short run. This occurs because of inefficiencies, inappropriate policies, insufficient funds, weak state capacity, and low political and societal will. Yet in the long run this growth will improve environmental conditions (once per capita income reaches a high enough level).

Research on the relationship between pollution intensity and per capita income by economists Gene Grossman and Alan Krueger and the World Bank in the early 1990s sparked a great deal of interest in the environmental Kuznets curve.[6] This research shows that as a country develops and industrializes, emissions of certain pollutants—such as sulfur oxides, suspended particulates (smog), and lead—increase along with GNP growth. In other words, growth and pollution are *coupled*. Pollution rises mainly because of an emphasis on industrialization, growth, and income generation, at the expense of pollution controls. Yet, when per capita income gets high enough (in the past between US$5,000 and US$8,000), the link *decouples* as economic changes, societal pressure, and state capacity converge to raise environmental standards. At that point emissions of these pollutants begin to fall (see figure 4.6 for an example of a hypothetical EKC).

A classic case of this relationship is Japan. After World War II, economic growth soared in Japan, resulting in an economic growth rate of 10.9 percent from 1950 to 1955, 8.7 percent from 1955 to 1960, 9.7 percent from 1960 to 1965, and 12.2 percent from 1965 to 1970.[7] Such rapid growth left Japan severely polluted by the 1960s.[8] The notorious mercury-poisoning case in the city of Minamata illustrates these appalling conditions. In the 1950s and 1960s many people in and around Minamata contracted a mysterious neurological disease. The cause, scientists eventually discovered, was exposure to methylmercury compound, which the Chisso industrial plant had discharged in factory effluent upriver from Minamata. The mercury had worked its way up the food chain until humans ingested it in seafood. Public outcry over the

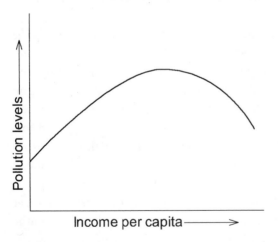

Figure 4.6
The environmental Kuznets curve.

Minamata disease contributed to rising citizen protests in Japan in the 1960s and 1970s over the environmental consequences of rapid industrial growth. In a relatively short time government and business cooperated to improve environmental policy and practice. By the end of the 1970s, Japan's environmental regulations had become among the most stringent in the world, and the environment was already much cleaner.[9] Today, given Japan's population density and small land mass, it is arguably one of the world's best-managed environments.

Most of the research on the long-term correlation between economic growth and improvements to environmental quality has concentrated on pollutants. Some studies have found, in addition, that the EKC holds for certain natural resources—such as a study of deforestation and national income in Asia, Latin America, and Africa.[10] The World Bank has also determined that the proportion of the urban population without adequate sanitation and safe water supplies drops as income rises in a cross-section of countries.[11] The World Bank qualifies these findings by adding that with respect to some indicators—such as municipal waste and carbon dioxide emissions—problems rise along with income. These caveats, however, have in no way undermined the belief among market liberals and many institutionalists that growth is good for overall environmental conditions.

Why do pollution levels and other types of environmental problems begin to decline as income rises? Intuitively one might think that because production and consumption rise along with income, pollution levels should rise too. Yet significant counterforces account for the inverted-U relationship between growth and environmental quality.[12] As an economy grows and people become richer, structural shifts can occur in the economy, such as a shift toward service- and information-based industries, which are less resource intensive and less polluting. At the same time societal concern over environmental issues (especially health problems) tends to rise as wealth brings education, communication technologies, and higher expectations. People begin to demand stronger government regulations to protect the environment, and are willing to pay the taxes to see these brought into being. Markets and governments push firms to develop pollution-control technologies, thus reducing emissions. There is also greater demand from consumers for "green" products as disposable income rises.

Most market liberals and institutionalists who support the EKC hypothesis do not, however, argue that countries should blindly pursue economic growth without environmental policies and only clean up after this becomes affordable. The EKC simply points to the value of economic growth for long-term environmental management. Growth itself, while it can create problems at certain points along a country's growth trajectory, does not always have to do so, especially if environmental measures are put into place early on.[13] Policymakers aware of this relationship can strive to "tunnel" through the EKC and reach higher levels of growth without doing any more harm than necessary to the environment.[14] Then economic growth in later stages can sustain an even cleaner environment. Market liberals also advocate market-oriented measures to promote environmental efficiencies to reduce the size of the peak in the curve, in particular clarification of property rights and reduction of market distortions (discussed later). Institutionalists focus much more on the need for international cooperation to help developing countries obtain necessary technologies to enable them to reduce the environmental impact of their economic growth. With such measures, growth can be turned into a positive force. On the other hand, both market liberals and institutionalists argue, slow growth or lack of growth can result in considerable environmental damage—possibly even more than rapid growth can.

Slow Growth and Lack of Growth

Slow growth, according to market liberals and institutionalists, does not allow for a future with better environmental conditions—just more of the same, or in some instances, cycles of increasingly weak economic growth and environmental collapses. In rich countries it limits government and corporate investments in cleaner technologies. It is an even greater problem for developing countries, because it keeps them on the wrong side of the critical point on the EKC where environmental improvements begin to occur naturally (and therefore efforts to reduce poverty tend to override environmental spending).

Slow growth reinforces the vicious cycle between poverty and environmental degradation, often the most serious barrier to achieving sustainability in the developing world.[15] Market liberals in particular focus on the environmental impacts of poverty. Institutionalists put more stress on how population growth interacts with poverty and environmental degradation (and call for better institutions to support family planning). The nexus between poverty, population growth, and environmental degradation produces a self-reinforcing downward spiral from which it is difficult to break out.

In both versions of this vicious cycle, the poor are seen as agents as well as victims of environmental degradation. They often have no choice but to overuse the land and overharvest resources around them for food and shelter. They must act to satisfy short-term needs even if this works against their long-term interests. The poor, then, are forced into decisions with short time horizons. How can we expect them to preserve natural resources for future generations?[16]

It is common for market liberals and institutionalists to link poverty with environmental degradation when analyzing rural and agricultural settlements in developing countries. The UN Development Programme (UNDP) calculates that "at least 500 million of the world's poorest people live in ecologically marginal areas," and many poor people in the developing world farm ecologically fragile land.[17] Often the poor have few options except to cultivate unsuitable land, which leads to soil erosion, land degradation, and deforestation. Growing populations in these areas reduce fallow periods as well, quickening the pace of soil degradation and deforestation.[18] Often the children of farmers struggle to get increasing yields on ever-smaller patches of land. Sometimes the situation is made worse because powerful elites (perhaps descendants of colonizers) control the most productive land. Sometimes it is made worse as settlers leave overpopulated areas to occupy "empty" land (usually small numbers of indigenous peoples actually reside in these areas). Some governments have even promoted such disastrous internal migration. Indonesia under former President Suharto (1965/7–1998), for example, implemented a Transmigration Program to move poor residents from the overcrowded island of Java to the outer islands (such as Kalimantan). Few of these people had the knowledge to farm these ecologically fragile outer islands. Many agricultural sites failed to produce adequate yields. The sites also upset the natural balance of local forest ecosystems, contributing to soil erosion as well as exposing and drying out large areas of peat moss. Appalling fires have swept through these areas (fueled in part by forests logged and degraded by firms linked to Suharto's family and cronies). The worst so far were in 1997–1998, when fires lit by plantation firms to clear land cheaply swept out of control and scorched 5–10 million hectares of land (including 1 million hectares of peat). The peat fires alone released 150 million tons of carbon into the atmosphere.[19] Klaus Topfer, then executive director of the UN Environment Programme (UNEP), described these fires "as one of the worst environmental disasters of the last decade of the century."[20]

Market liberals and institutionalists, then, point to the poor as critical forces contributing to environmental problems. They also recognize, however, that the poor, without the means of avoiding the impacts, suffer the most from environmental degradation. The poor are more vulnerable to bad water and poor sanitation. They often live closest to the sources of pollution in urban centers. And they rely on types of fuel, such as wood and charcoal, that produce poor air quality in their own homes. This all feeds back to make the poor even poorer as productivity declines—a situation that in turn further accelerates the destructive impact of the poor on the environment.[21] In this way poverty and environmental degradation are mutually reinforcing, as depicted in figure 4.7. A World Bank report on economic adjustment in Africa illustrates this poverty-population-environment analysis:

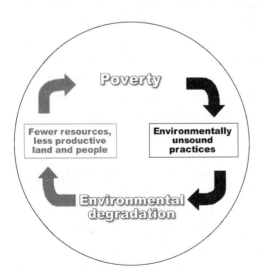

Figure 4.7
The vicious cycle of poverty and environmental degradation.

The poor are both victims and the agents of damage to the environment. Because the poor—especially poor women—tend to have access only to the more environmentally fragile resources, they often suffer high productivity declines because of soil degradation or the loss of tree cover. And because they are poor, they may have little recourse but to extract what they can from the resources available to them. The high fertility rates of poor households further strain the natural resource base.[22]

The poverty-environment spiral, argue market liberals and institutionalists, occurs for many reasons. Market liberals focus on inadequate economic growth and insecure property rights as the key causes of poverty, while institutionalists tend to stress the role of weak government services. Weak economic growth in part arises because of insufficient domestic integration into global markets. Insecure property rights, especially in cases where arrangements for commonly-held property (common property regimes, or CPRs) break down into chaos and open access, create strong incentives for the poor to overexploit local resources. Insecure property rights also leave the poor without land title and therefore less able to obtain credit in times of need. Women in particular are often shut out of credit systems because of weak property rights in many poor countries. Inadequate government services, such as education, are also unevenly distributed among the poor, with women often being denied equal access—conditions that perpetuate unsustainable resource use.[23] Multilateral and bilateral aid programs to promote "women in development" are now common. An underlying purpose of these programs is generally to promote sustainable development.

Both market liberals and institutionalists emphasize the need to alleviate poverty to end the spiral of poverty and environmental degradation. For market liberals this requires the integration of poor countries into the global marketplace—that is, countries need to liberalize trade, deregulate "overmanaged" industries, and improve the investment climate for transnational firms. To carry out these measures, governments must accept global standards (private property, customs, banking, taxation, environmental regulations, and so on) as well as eliminate corruption and ineptitude. Only then can these countries reduce poverty levels and move toward a future with better environmental conditions. Institutionalists agree to some extent with this prescription, although they place even more emphasis on the need to build the capacity of local and national governments to manage the integration into the global economy. Institutionalists focus more, too, on population and gender measures, such as improving access to credit for women and making more effective family planning available.

Market Distortions, Inefficiencies, and Weak Institutions

Market liberals in particular see market distortions as a serious problem for the natural environment. Market distortions are perversions of free markets causing a social or environmental outcome that is suboptimal. These distortions create inefficient use of resources and sinks, with negative outcomes for both society and the environment.

Government subsidies to promote economic growth in certain sectors or to diversify the economy are one of the most common examples of market distortions. Governments may provide financial support (in the form of payments, tax breaks, and so on) to certain sectors as a way to encourage investment and improve their international competitiveness.[24] Many industrialized countries, for example, subsidize the mining and logging sectors by charging what many consider to be below-market prices for the rights to extract these resources. The United States and Europe Union countries also use a variety of support instruments to subsidize their agricultural sectors, such as direct payments and cheap credit, to encourage food production and exports. Developing countries often subsidize certain sectors as well, most commonly as a strategy to move away from an economy with a heavy reliance on a few unprocessed natural resource exports (like bananas, gold, or logs) and toward one where domestic firms process these natural resources before export (which adds economic value and creates jobs). Energy subsidies are common too in the developing world as a means of promoting this kind of industrialization.

Market liberals argue that even though the ultimate goal of such subsidies may seem sound and sensible, they ultimately contribute to wasteful and inefficient domestic industries that overconsume local resources and pollute local ecosystems. Artificially low prices of timber, minerals, oil, or agricultural chemicals encourage overuse and waste, leading to environmental degradation and inefficient markets. Subsidies, too, can undercut incentives to upgrade environmental technologies. This is common, for example, with fuel subsidies that weaken incentives to use more fuel-efficient technologies. While market liberals most often make this point, it should be noted that bioenvironmentalists, social greens, and institutionalists all tend to agree that governments should eliminate subsidies with negative environmental implications. The market liberals, though, are against most forms of subsidies on the grounds that they are inefficient, whereas the other groups are more open to them as tools—for example, to promote cleaner production or to pursue other goals.

Both market liberals and institutionalists have begun to promote the notion of sustainable consumption to encapsulate the environmental drawbacks of inefficient production and consumption. This concept is often formulated so as to incorporate both the dematerialization of production and the optimization of consumption. Some refer to this as "ecoefficiency"[25] and "cleaner production," concepts that involve decoupling higher production and consumption levels from their negative impact on the natural environment. The idea is that such practices are not only good for the environment, but also more efficient and thus improve the profitability of firms as well. Business, here, is seen as an environmental leader, as the pursuit of profits becomes the pursuit of more efficient use of the environment. The notion of sustainable consumption seems to be gaining strength. The OECD, for example, now has a research program on sustainable consumption.[26]

How, then, does the global community achieve sustainable consumption? Market liberals argue for market-based measures that encourage cleaner technologies, such as sensible taxes, transparent and equitable user fees, and tradable permits. Also, governments need to eliminate subsidies that encourage individuals and firms to use outdated and unclean technologies. If such changes take place, market liberals argue, the free market is the best way to develop and disseminate green technologies, via global corporations.

Institutionalists argue that greater global cooperation—for example, global standardization of reasonable environmental laws, environmental certification programs, government support, and global environmental regimes—is necessary to assist with the spread of cleaner production and sustainable consumption. The global community needs to raise awareness so more individuals begin to demand greener products. International institutions such as the UNEP can help promote such sustainable consumption. So can international agreements that help to reduce the use

and impact of certain industrial and natural resource inputs. Governments can provide financial support for research, development, and dis-
semination of green technologies. Finally, green aid can help to transfer
cleaner technologies to developing countries.[27]

Critiques: Bioenvironmentalists and Social Greens

Bioenvironmentalists and social greens are highly critical of mainstream
neoclassical economics, particularly the emphasis on "economic growth"
and GNP/GDP as measures of human well-being. Growth, in fact, is not
a source of better environmental conditions, but a source of environmental degradation at all income levels. Academics known as *ecological
economists* have developed some of the most compelling critiques. Many
of these arguments began to emerge in the late 1960s and early 1970s,
and were part of a broader global questioning of the environmental
implications of the global political economy (see chapter 3). Ecological
economists, who tend to fall into the category of either bioenvironmentalist or social green, reject many of the assumptions and principles of
the neoclassical economic model and propose an alternative model.
Because it is based on different assumptions, ecological economics is very
different from the "environmental economics" discussed earlier, because
the latter incorporates environmental concerns into what is more or
less a neoclassical economic framework. Perhaps the most significant
assumption of ecological economics is that the environment is not a
subsystem of the macroeconomy, but rather that the economy is a
subsystem of the natural environment or ecosystem (see figure 4.8).

Framing the economy-environment relationship in this way has profound implications for how ecological economists analyze the causes of,
and solutions for, global environmental change. The model highlights
the biological "limits" to economic growth—that is, the view that the
earth is a finite resource. How soon will we reach these limits? For some
ecological economists we are already beyond them. For others we are
plummeting closer by the day. We now examine these arguments more
closely.

Problems with GDP/GNP as a Measure of Development

Social greens and bioenvironmentalists both argue that GNP and GDP
are generally poor measures of development and of the human condi-

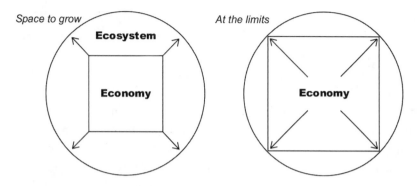

Figure 4.8
The economy as a subsystem of the ecosystem. When viewed as a subsystem of the ecosystem, economic growth eventually comes up against the earth's natural limits on resources and sinks. *Source*: Adapted from Daly 1996, 49.

tion. GNP and GDP only count things that go through the market—that is, goods or services that involve an exchange of money. A great deal of work that enhances the quality of life does not go through the market, however, and thus is not counted in GNP or GDP figures. This includes, for example, work that is done in the home, such as housekeeping and childcare, and volunteer community work. Social greens in particular stress that much of this work is done by women, and that GNP and GDP figures seriously undervalue women's work.[28] The way national wealth is measured also undervalues subsistence farming, even though in some parts of the world this is the primary source of food. Linked to this critique is the fact that markets often ignore or underprice environmental goods and services, like fresh sources of water or clean air. These either do not enter the calculation of GNP/GDP at accurate prices, or do not enter it at all.[29]

At the same time, GNP and GDP do count things that we may consider "negatives" for the human condition, such as industrial pollution and resource extraction. Some argue that GNP and GDP should subtract such costs.[30] Similarly, these measures count spending on services to deal with human catastrophes,[31] like cleaning up after an environmental disaster (such as the Exxon Valdez oil spill) or terrorist attack (such as the World Trade Center attack on September 11, 2001), simply because these efforts do go through the market. GNP also tells us little about things that are not easy to quantify, such as the state of the natural environment or human rights. Nor does it tell us much about income distribution. While most mainstream economists do admit that GNP is a simplified and far-from-perfect measure of human well-being, they nonetheless continue to use it.[32]

The UNDP's human development index (HDI) tries to address some of these critiques by incorporating education and health indicators, but because the main HDI index does not include environmental quality and still includes economic growth as a major component, many social greens and bioenvironmentalists see it as an inadequate measure of development. They would like to see governments react to changes in ecological measures with the same vigor as with economic and social ones. Yet environmental policymaking has no comparable measures today. Many scholars and research institutes have proposed ecological indicators in the last decade or so, but so far none have gained much ground among policymakers. (Chapter 8 discusses these indicators in more detail.)

Limited Natural Resources and Sinks

We noted previously that neoclassical economists do not see economies as part of ecosystems, but rather ecosystems as part of the economy. Because they view the world this way, they do not see resources as being necessarily finite in an economic sense, a position reinforced by the steady decline in the prices of natural resources in real terms (box 4.2 summarizes the now-notorious bet between Julian Simon and Paul Ehrlich over the scarcity and prices of the globe's natural resources). These assumptions have been challenged from many angles, including by ecological economists who have drawn on the laws of thermodynamics (in particular the "entropy law") to illuminate the link between economies and the ecosystems (see box 4.3 and figure 4.9).

These thinkers, many of whom closely adhere to the bioenvironmentalist perspective, use the laws of thermodynamics to show that the notion of perpetual growth, promoted by market liberals, is erroneous because unlimited growth is physically impossible. By linking economics to the physical sciences, these thinkers have opened an entirely new way of viewing economic processes. For them, the economy is bound by the energy and matter physically available to fuel the production of consumer goods, and to absorb the waste products associated with those goods. In other words, there is a physical basis for scarcity, not just an economic one.[33] They argue that unless we recognize the very tangible

Box 4.2
The Simon-Ehrlich wager

In 1980 the economist Julian Simon and the ecologist Paul Ehrlich made a 10-year bet on the direction of natural resource prices. Ehrlich chose five metals—chromium, copper, nickel, tin, and tungsten—worth US$1,000 (US$200 of each metal). The wager was simple: If, after adjusting for inflation, the market price of these metals in 1990 was more than US$1,000, Ehrlich would win (because this would demonstrate growing scarcity). If the total was less than US$1,000, Simon would win, because this would confirm the economic infinity of the resources, since new substitutes and technologies kept prices low. The loser had to pay the winner the total difference in the 1990 inflation-adjusted prices. Ehrlich seemed confident at the beginning: "Simon is wrong about the economics of mineral resources. . . . The trough-like pattern long predicted for mineral resource prices has now shown up. . . . I and my colleagues, John P. Holdren . . . and John Harte [Berkeley physicists] . . . jointly accept Simon's astonishing offer before other greedy people jump in."[1]

Simon turned out to be the more astute gambler, however, and by 1990 the market prices of all five metals were below the 1980 levels (in real terms). The price of the nickel and copper were only down to US$193 and US$163 respectively, but chrome had fallen to US$120, tungsten to US$86, and tin to US$56. Ehrlich and his colleagues paid Simon a sum of US$576.07. Reporter John Tierney asked Ehrlich: "Is there a lesson here for the future?"

"Absolutely not," Ehrlich answered. "I still think the price of those metals will go up eventually, but that's a minor point. The resource that worries me most is the declining capacity of our planet to buffer itself against human impacts. Look at the new problems that have come up, the ozone hole, acid rain, global warming. . . . If we get climate change and let the ecological systems keep running downhill, we could have a gigantic population crash."

When Tierney told Simon of Ehrlich's response, he was quick to retort: "So Ehrlich is talking about a population crash. That sounds an even better way to make money. I'll give him heavy odds on that one."[2]

1. Ehrlich 1981, 46.
2. Tierney 1990. Also see Simon 1996; Ehrlich and Ehrlich 1996.

Box 4.3
Entropy and ecological economics

The pioneer ecological economist Nicholas Georgescu-Roegen first outlined the relevance of entropy to the field of economics.[1] Entropy is the amount of usable energy or matter in an isolated system. High entropy means the energy or matter is not very useful, whereas low-entropy energy or matter is useful and can perform work. The second law of thermodynamics, also known as the entropy law, states that the amount of usable energy and matter decreases in an isolated system as work is performed. In other words, when energy or matter is transformed to make a product, its entropy increases, making it less useful.

The earth provides us with some forms of low-entropy energy—that is, energy that is easily usable by humans to perform work. Ecological economists view the earth as a closed system (i.e., one that energy flows through while matter cycles within), and argue that the functioning of the current global economy neglects the implications of the second law of thermodynamics.[2] In this view of economics, nature provides the earth with a flow of low-entropy energy (in the form of solar energy) and a set stock of low-entropy nonrenewable energy (e.g., from terrestrial sources, such as fossil fuels). Modern economies tend to exhaust the terrestrial stock of low-entropy energy. When we use this natural resource–based energy to make products, or burn it as fuel, referred to as "throughput," we end up with less useful, high-entropy energy, such as carbon dioxide, that is disbursed into the atmosphere.

Ecological economists like Herman Daly argue, following Georgescu-Roegen, that the more we use up low-entropy energy to support the functioning of our modern economy, the more high-entropy waste energy we are left with. The sun still provides a limited flow of low-entropy energy and will do so for the foreseeable future. Humans have more control over the flow rates of the use of low-entropy terrestrial stocks of energy. Daly argues that we are using these terrestrial stocks of low-entropy energy far too quickly, and that once these run out it will be extremely difficult to reorganize the economy to run on the available flow of solar energy. He uses a similar line of reasoning to argue that we are using our terrestrial stock of low-entropy matter (nonrenewable resources, such as minerals) and our flow of low-entropy matter (renewable resources, such as timber, that rely on the sun for replenishment) at a rate that is out of line with the earth's capacity to replace them.[3] In other words, there is too much physical "throughput" in the economy. The importance of entropy is illustrated in the entropy hourglass, illustrated in figure 4.9.

1. Georgescu-Roegen 1971.
2. Georgescu-Roegen 1993; Costanza et al. 1997, 56–59.
3. Daly 1996, 29–30, 193–98.

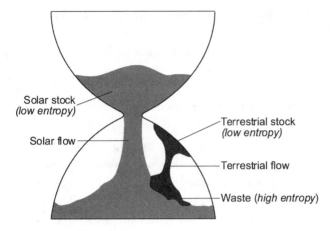

Figure 4.9
Entropy hourglass. *Source*: Georgescu-Roegen, as depicted in Daly 1996, 29.

and real physical limits of nature and learn to live within them, we will overstep the ecological bounds of the planet. Some fear we are already beyond those bounds.[34]

Because the global political economy fails to account for these limits, it fails to send signals when resources are being used in an "unsustainable" manner. Thus, directly counter to Simon's argument discussed in the first part of this chapter, for ecological economists prices are not a good determinant of the availability of resources. For them, the discovery of new resources will not solve this inherent defect in the global political economy. All it does is to substitute the use of one limited resource for another. Terrestrial stocks of natural resources are still finite, and the flow of solar energy and renewable resources is limited at any given time as well. Such limits imply that market failures are a much greater problem than most economists acknowledge—that is, negative externalities that result in pollution and resource exhaustion are inherent in economic growth, and are not a rare occurrence that a few market-oriented mechanisms can fix.[35]

Market liberals and institutionalists have not spent a great deal of time directly engaging with these ideas. Ecological economist Herman Daly, whose work is devoted to promoting the idea of the physical limits of economic growth, worked as a senior economist in the World Bank's Environment Department from the late 1980s to the mid-1990s. Though the World Bank focused much more on environmental issues during his tenure (see chapter 7), Daly eventually left that institution, many speculating that in part it was because the Bank was not taking his ideas seriously enough.[36]

Critique of the Environmental Kuznets Curve

Bioenvironmentalists and social greens are both highly critical of the notion that growth eventually improves environmental conditions, as illustrated by the EKC. Many of these thinkers have raised concerns with respect to the methodology used to determine the EKC.[37] Many argue, too, that the connection made between growth and pollution levels as a general indicator of environmental quality is too simplistic, and therefore that relying on the EKC to guide policy is inappropriate.

There are many specific criticisms. First, the research on environmental Kuznets curves only applies to a limited range of pollutants and resources. It does not apply, for example, to CO_2 emissions, which continue to rise with economic growth.[38] Moreover, it is insufficient to simply plot a decline in the emissions of specific pollutants as income rises. It is entirely possible that higher emissions of other pollutants in the same economy may simply be accounting for this decline. For instance, sulfur dioxide emissions may decline with a switch away from coal-fired power plants, but other types of emissions may rise, such as nuclear waste from nuclear-fueled plants. If the current population does not bear the environmental costs of such emissions, as is the case with nuclear waste, it is less likely that a rise in income will result in lower emissions of those pollutants.[39]

Similarly, a decline of particular pollutants in one country may occur because these have been shifted to other, often poorer countries. Japan, for example, partly in an effort to improve domestic environmental conditions, has transferred dirty industries to the developing countries in Asia and the Pacific.[40] Much of the initial data on the EKC was from OECD countries. It is not clear what would happen to the relationship between income and environmental quality if we took a global perspective. We do know, however, that as certain types of environmentally harmful production decline in rich countries, they have tended to increase in poor countries.[41] Economic globalization, critics of the EKC argue, makes pollution displacement to poorer countries even more likely.[42] Accounting for displacement is critical. Global pollution levels may continue to rise even as levels fall in some wealthier countries.[43]

Critics of the EKC note further that it ignores the feedback from environmental degradation. In other words, it assumes that environmental degradation does not harm future growth.[44] Yet some forms of pollution are cumulative (CO_2, CFCs, and nuclear waste) and some degradation is permanent (biodiversity and species loss). This sort of environmental harm can and does create negative feedback, holding back future growth. If this is the case, it may be possible for a country to get "stuck" along the curve, never experiencing the decline in pollution that the EKC predicts.[45]

Poverty as a Scapegoat for Environmental Degradation

Social greens and bioenvironmentalists do not deny that poverty and environmental degradation are linked. However, they see that linkage as being far more complex than market liberals and institutionalists portray.[46] They argue that if we look more deeply and differentiate between various types of poverty, it becomes clear that this relationship is far from straightforward and varies with the type of poverty a particular community is experiencing.[47] While many bioenvironmentalists acknowledge this important point, it is the social greens who have focused most on debunking the idea that the poor are responsible for environmental degradation. Rather than asking simply why the poor degrade the environment, social greens in particular ask why the poor are poor in the first place.

For social greens, it is not poverty itself that causes environmental degradation. Instead, it is the inequalities of the global economy that marginalize people and put them into precarious situations that sometimes, but not always, result in environmentally damaging outcomes.[48] For these thinkers, when there is significant environmental damage among the poor, it is a result of development itself, rather than the lack of it.[49] For social greens, then, it is not the poor farmer that is responsible for the loss of biodiversity in the developing world; it is the spread of industrial agriculture. Small traditional farmers plant many varieties of seeds in their fields, whereas modern farming practices tend to rely on monocultures, creating much more vulnerability, risk, and soil degradation.[50]

It is impossible, then, to fully understand the poverty-environment linkage until we first put poverty into its broader context, and ask how national and global policies and economic relationships impact poverty at the local level. Social greens in particular argue that we need to look beneath the surface in each specific case of environmental degradation amidst poverty.[51] In other words, rather than blaming the poor, we must examine the forces that pushed the poor onto marginal lands in the first place, and ask whether these are the forces responsible for environmental degradation.[52]

Often, social greens argue, one of the main forces that push the poor onto marginal lands is the process of "enclosure"—or privatization of lands once held in common-property regimes (CPRs). CPRs can be quite sustainable property arrangements, drawing on local community knowledge and longstanding mechanisms of control.[53] These thinkers frequently point out that the so-called tragedy of the commons described by the bioenvironmentalist Garrett Hardin is in fact a tragedy of open-access regimes, which are quite different from commons regimes.[54] They argue that CPRs are often mistaken for open-access systems, and unnecessarily attacked as "unenvironmental."

Social greens see enclosure as part of a long historical process whereby more powerful actors, including states and corporations, have taken over control of the commons. The powerful actors in the global economy in particular have tended to confiscate the property rights of the poor (especially women and indigenous peoples), pushing them onto marginal lands and into a fragile existence. Minorities whose lands have been seized have then often been the target of state violence to repress their resistance.[55]

Many social greens view enclosure as an outgrowth of the global spread of capitalism. This process has enabled transnational corporations or local elites to occupy the best lands for timber or export-crop production, which in turn creates pockets of poverty with severe environmental problems (like soil erosion and falling agricultural yields). Indigenous and women's knowledge is the latest enclosure of the commons. In this case, powerful actors are claiming peoples' knowledge of their local environments—for example, the medicinal value of seeds and plants—as their private property in the form of patents.[56] Activist-scholar Vandana Shiva, for instance, argues forcefully that commoditizing and privatizing such knowledge leaves many of the world's poor even poorer. Few, if any, locals benefit from this process; in some cases, locals can even lose the right to "use" their own knowledge.[57]

The agricultural practices of the poor, many social greens note too, are (or once were) fully sustainable. One example is shifting cultivation, where farmers move from one plot to another after several years of cultivation, often burning the new plots to clear them. Market liberals, mirroring colonial attitudes in developing countries, conclude that this type of farming creates excessive deforestation and soil degradation in poor countries. Institutionalists and some bioenvironmentalists add that population pressures often exacerbate this problem, or at least make this once-sustainable practice unsustainable. Social greens stress, however, that this type of cultivation is still generally more ecological than industrial monoculture agriculture, resulting in richer soil for the crops with fewer weeds, and acting to prevent more serious and devastating fires.[58] Blaming the poor and population pressures in these cases might be easy, because their global voice is weak, but for social greens it is ethically *and* ecologically wrong.

A further point made, in particular in the social green literature on the poverty-and-environment linkage, is that the poor in many cases are environmental activists and environmental protectors.[59] The poor are often women and indigenous peoples who have been forced off their ancestral lands by state and global political and economic processes. They may lack material wealth, but they have a rich wealth in local knowledge—much of which is centered on practices that preserve the environment in the long run. Their marginalization, however, often deprives them of the opportunities to follow these practices.[60]

There has in fact been a rise in ecological resistance movements particularly in the developing world, mainly made up of the poor who are passionate about the environment. In many cases these activists are women whose access to resources is most threatened by industrial development. The movements in India against the Narmada Dam (discussed in chapter 7) and the Chipko movement, the Greenbelt Movement in Kenya, and the rubber tappers' movement in Brazil (discussed in chapter 3) are some important examples of this type of grassroots resistance.[61] Such activists see themselves as fighting against outside forces that contribute to environmental degradation. Political scientist Robin Broad argues that when the poor rely on the environment to survive, when there is a sense of permanence on their land, and when civil society is highly organized, environmental activism is likely to be strong among the poor. Those without land rights, or who have been forced off their own ancestral land and onto other land, and where political regimes are repressive, are less likely to become environmental activists.[62]

Policies that flow from this analysis center on the notion of regaining community control of the commons. For this reason most social greens are highly skeptical of the market-liberal view that we need to have clearer property rights, and in particular private property and land title, for the poor. Social greens call for a reordering of the global economy toward a more local one. As stressed by *The Ecologist*, this involves "pushing for an erosion of enclosers' power, so that capital flows around the world can be reduced, local control increased, consumption cut, and markets limited."[63]

Wealth and Overconsumption

Bioenvironmentalists and social greens see environmental problems associated with extremes of wealth and poverty as closely related to one another. They argue that the globalization of the world economy is one of the main drivers leading to ever more skewed income and consumption inequalities. They are critical of the market-liberal and institutionalist view that more globalization will raise all boats, because to them the finiteness of resources dictates a zero-sum equation. Wolfgang Sachs, a leading social green scholar and activist, puts it like this: "In a closed space with finite resources the underconsumption of one party is the necessary condition for the overconsumption of the other party. . . . The rising tide, before lifting all the boats, is likely to burst through the banks."[64]

Bioenvironmentalists and social greens both argue that growing consumption, itself tightly linked to economic growth, is devastating the planet. The world has seen a fourfold increase in energy consumption since 1950.[65] Consumption doubled in real terms from 1973 to 1998 (global consumption expenditures were US$24 trillion in 1998).[66] Manufactured products make up a growing proportion of this rising consumption. Rich countries have much higher rates of per capita private consumption (around US$16,000 in 1995) than developing countries do (only US$300 in much of the developing world). Africa's consumption has actually declined in the past 20 years, while it has risen just about everywhere else. Industrialized countries, though only accounting for 15 percent of global population, are responsible for 76 percent of global consumption expenditure.[67] This disparity remains even though consumption is rising faster in developing countries than in industrialized countries.

This skewed distribution of consumption also has unequal impacts on the use of natural resources—both nonrenewable and renewable. The United States alone, with only 5 percent of the world's population, consumes 30 percent of the world's resources.[68] The rich primarily benefit from this use of resources, yet the environmental costs are spread widely. There are, for example, severe problems with oil drilling to sustain the global industrial system. As reserves dwindle, the net is cast ever wider, and more and more investment is making its way into the developing world (see chapter 6 for further details). The global trade in timber and minerals also primarily affects the developing world, while most of the resources end up in industrialized countries.

One way of demonstrating this inequality in consumption and its impact on resources and sinks is ecological-footprint analysis. William Rees and Mathis Wackernagel developed this approach as an accounting tool to measure the sustainability of average lifestyles. It translates human consumption of food, housing, transportation, and consumer goods and services into hectares of productive land. This allows for individuals and states to compare their ecological footprint on the planet—a vivid way for a person to imagine his or her ecological impact on the sustainability of life on earth.[69] Ecological footprints vary enormously by region. For example, the ecological footprint of an average person in Africa is 1.36 (e.g., for Burundi it is 0.48; Ethiopia, 0.78; Mozambique, 0.47), while the average ecological footprint of an American is 9.7, a Canadian 8.84, and an Australian 7.58.[70] It would require four planet earths for everyone in the world to reach present U.S. levels of consumption.[71]

The centrality of the concept of inequality to social green and bioenvironmentalist thought goes back to the environmental movement in the late 1960s and early 1970s. As chapter 3 discussed, at Stockholm the environmental movement and developing countries pushed the theme of inequality. More recently, social greens in particular have pinpointed globalization as a key driver of inequality, which in turn exacerbates the process of environmental degradation. The antiglobalization movement, which includes many social green thinkers, has also taken up this line of argumentation. This critique of globalization has been bolstered by academic studies on the left that pinpoint globalization as a major contributor to global economic inequality.[72]

Bioenvironmentalists tend to emphasize both overconsumption in the rich world and population growth in the developing world, because both contribute to growing global consumption.[73] They are particularly concerned with consumption rates *per person*, and on this front the industrialized world tends to take the most criticism. According to leading environmental analyst Norman Myers, "The main responsibility for grand scale consumption lies with the rich; not only do they consume a disproportionate quantity of resources, but they have more opportunity to modify their lifestyles than people struggling to raise themselves out of poverty."[74]

This line of analysis is inspired by the IPAT formula outlined by Paul Ehrlich and John Holdren in the early 1970s. This formula, $I = P \times A \times T$, postulates that environmental *impact* is a product of *population* multiplied by *affluence* (consumption per person) and *technology* (which affects the amount of resources and wastes required to produce each unit of consumption). Though the formula is simple, Ehrlich and Holdren felt that it got to the root of the environmental problem, and showed how inextricable each factor was from the others. While in the early 1970s most bioenvironmentalists focused on population growth in the developing world as a core stress on the global ecosystem, today many worry more that increasing global trade will harmonize consumption rates upward, with developing countries eventually consuming at the rate of the developed countries (all within the context of an increasing global population).[75]

Most social greens sharply criticize the inclusion of population in the discussion of consumption, in favor of an analysis more squarely focused on the rich world's consumption patterns.[76] They stress the political use of population in much of the mainstream analysis. As *The Ecologist* notes:

The "population issue" is being used politically to control groups of people who are considered a threat by those who have benefited from enclosure. Moreover, "population control" is being used to enclose women further: it takes away their power to determine how many children they have and to choose a method of regulating births which they control and which does not threaten their health.[77]

For them, then, it is dangerous to focus on the question of population, particularly in the developing world, because it disempowers women, who make up the majority of the world's poor. Instead, they focus squarely on the rich world's consumption patterns, and also actively defend the right of all women to have control over their own bodies.[78]

On the question of consumption, both social greens and bioenvironmentalists point to what they call the problem of "overconsumption." As the rich get richer and consume more, they inflict more damage on the natural environment than is necessary. This is because much of their consumption is based on wants and not needs. Luxury goods that are far beyond what is necessary are seen to be proliferating. For many in the rich industrialized world, shopping has become a pastime rather than a necessity, creating or exacerbating a culture of consumption.[79] The growing presence of sports utility vehicles (SUVs) in urban centers in North America over the past decade, just as incomes were rising, is a case in point. Often equipped with four-wheel drive and truck engines for traveling on rough terrain, they are not really necessary for city driving conditions.

This overconsumption by the rich is exhausting energy supplies and contributing to "overpollution" and "overgeneration of wastes." This is true of both corporate and individual consumption.[80] People now buy more things and throw more things away than in the past. This arises partly because products like computers become obsolete in just a few years, partly because engineers intentionally design some products to last for a short time, and partly because an increasing number of people now "enjoy" replacing their possessions (a trend reinforced by advertising).

Consumption in rich countries has continued to rise at an increasing pace over the last 200 years. So far, contend many social greens and bioenvironmentalists, these countries have managed to skirt the natural limits of resources by acquiring resources from other parts of the world.[81] In the past this was done mainly through colonialism and migration to new lands. Today it is done through international trade, foreign direct investment, and foreign aid. A country's "ecological shadow" refers to its aggregate environmental impact in other parts of the world to feed domestic consumption and avoid domestic environmental problems.[82] Globalization is increasing the reach and size of ecological shadows, shifting the environmental burden from consumption far beyond the borders of rich countries, and allowing the rich to consume ever more without suffering any of the consequences of hitting ecological limits. The "bigness" of the global economy supports the lengthening of supply and disposal chains for consumer products. This "distances" consumption from extraction and production[83] and from its final resting place as waste,[84] making the environmental implications of consumption harder for consumers to comprehend and act on. One example is computers. The production and disposal of computers involve toxic metals and plastics. The hazardous part of computer production commonly occurs in developing countries. Developed countries also commonly export computer waste to developing countries for recycling.[85]

Why does this trend occur? Social greens argue in particular that the privileged position the rich countries hold in the current international political economy has enabled them to continue their high per capita consumption of natural resources by buying up the developing world's resources at cheap prices, which contributes to both poverty and environmental degradation. The ideology of growth and industrialization, part and parcel of capitalism, is blamed for contributing to the dominance of the industrialized countries and transnational corporations in an increasingly global economy in ways that weaken the power of the poor countries to resist the environmental degradation inflicted on them.[86] This situation is seen as part of a longer historical trend. In the past the rich countries extracted resources from colonies; today they do so through the process of economic globalization. Both eras are similar; involving uneven development and uneven distribution of environmental degradation, with the poor suffering the most.[87]

The wealthy, argue social greens and bioenvironmentalists, must curtail consumption. There is no other reasonable and fair option. Bioenvironmentalists stress the need to lower total global consumption. This will require the rich countries to lower consumption, but equally important for many bioenvironmentalists is the need for the poor countries to control population growth. In making this recommendation, they are not that far from the suggestions of institutionalists who stress sustainable consumption and population management, discussed in the first part of this chapter, even though they arrived at this point along a different path. Social greens emphasize the need to reduce consumption inequities—that is, lower consumption among the wealthy, so more are getting what they need (sufficiency), rather than a few getting what they want (gluttony). It is critical, too, that local communities incorporating the views of women and indigenous peoples maintain control of local resources and do not relinquish this control to transnational corporations. This will require not just more individual responsibility for environmental management, but also broader political and economic changes to reward less consumptive, more local economies.[88]

Conclusion

This chapter demonstrates the complexity of the ecological debates over the impacts of economic growth, poverty, and wealth. There is no doubt a major fault line on the question of economic growth, with a coalition of social greens and bioenvironmentalists on the critical side, and a coalition of market liberals and institutionalists on the other. A deeper look has revealed, however, that on other issues different coalitions have emerged. On a few points there is even common ground among all of the views.

On the main question of economic growth and whether it is positive or negative for the environment, market liberals and institutionalists both advocate a pro-growth strategy with the aim of reducing poverty and raising revenues to pay for environmental improvements. Market liberals stress that promotion of a free market is the best way to foster growth, because price signals should encourage innovation and greater efficiency in resource use. Alternatives to oil, for example, will develop when prices dictate, a process already underway with alternatives like solar and wind power. Market tools—such as ecotaxes, user fees, and tradable resource rights—can help to address distortions and market failures. This does not mean that all environmental regulations are bad, just that, whenever possible, market approaches tend to create a better long-term environmental-management structure. Institutionalists are much more cautious about leaving everything to the free market. They want to see policies and structures that help economies tunnel through the EKC—in other words, more active policies to better weather the inevitable pollution en route to the cleaner world of wealth and prosperity. One such policy is rooted in the idea of sustainable consumption.

A coalition of bioenvironmentalists and social greens on the other hand argue that economic growth directly harms the environment. It is not going to lift the poor out of poverty, but rather, given the current structure of the global economy, will merely exacerbate inequalities. Bioenvironmentalists and social greens call for a total rethink of consumption and growth and nature. They both agree that this requires that the physical "throughput" in the economy does not outpace the supply of renewable and nonrenewable natural resources. It is still possible to provide for a "good life" for present and future generations. But doing so will require slowing the physical aspects of "growth" based on resource use to sustainable levels. This does not necessarily mean "zero economic growth," but it would mean a radical cutback. Social greens take the point further by calling for a redistribution of global economic resources and a refocus on local-community development instead of emphasizing economic growth per se. For them, there are many ways to work toward this broader concept of development, including local economic trading systems, reinstatement of common property regimes in the developing world, and tight restrictions on transnational corporations and trade.

On other issues, such as the relationship of population to poverty and environmental degradation, we see a different convergence of arguments among the worldviews. Bioenvironmentalists remain particularly concerned with population growth, viewing it as intricately tied to consumption and economic growth. Institutionalists share many of these concerns, seeing population growth as tightly tied to the vicious cycle of poverty and environmental degradation. Thus, both feel that policy initiatives to promote sustainability should incorporate some sort of population-control measures. On the other hand, social greens view the inclusion of population in the discussion of consumption and growth as dangerous, because often it becomes a political tool to control women and the poor. For different reasons, market liberals also do not focus much on the links between population growth and the environment. For them, resource scarcity is not a pressing environmental issue. Therefore neither is the number of humans.

This examination of the debates over growth, poverty, and wealth shows, too, a convergence of all views on some points. This includes most prominently the idea of reducing environmentally harmful subsidies and the promotion of green taxes. While environmental economists (many of whom adhere to market-liberal or institutionalist views) have proposed many of these ideas, ecological economists (many of whom adhere to bioenvironmentalist or social green views) embrace them, too. The Green Party in many countries, for example, has called for a "tax shift" toward environmentally sound taxes.[89] The different groups we have been surveying have arrived at this policy prescription using different assumptions and arguments. Social greens and bioenvironmentalists, for example, are skeptical of growth in general and see the tax shift as a way to slow down especially ecologically destructive growth. Market liberals see growth as environmentally positive, and efficient growth, fostered by a reduction in subsidies and green taxes, as even more positive. Institutionalists see these tools as a way for institutions such as the state to enhance their capacity to promote environmentally sound growth.

This chapter has shown, then, that when it comes to these particular debates each worldview has its own unique take and the coalitions are not the same on all issues. The next chapter on trade further demonstrates the value of comparing the particular arguments and assumptions of each worldview.

5

Global Trade and the Environment

Is global trade saving the planet? Or destroying it? There are cogent arguments for both positions.[1] Market liberals see a free flow of goods around the world as a positive environmental force. From this perspective, global trade has many benefits. Free trade, through the logic of comparative advantage, creates global wealth and prosperity. It is efficient, avoiding waste and duplication. It also provides less industrialized regions of the world with advanced technologies and goods in exchange for relatively simple products (like minerals or timber). Global trade allows a country with a comparative ecological and labor advantage in, say, forest management, to develop a viable economy around timber exports. The ensuing societal wealth then enables citizens to purchase imports (like cars and computers), many of which would be well beyond the capacity of the country to manufacture (or, at best, terribly inefficient to do so). There is, moreover, little choice for poorer countries in an era of globalization, because restricting trade only leads to hardship and greater inequalities. Instead, states, indeed the entire global community, need to liberalize trade as much as possible. Free-trade institutions like the World Trade Organization (WTO) and regional trade agreements like the North American Free Trade Agreement (NAFTA) are therefore critical to the health of the planet, because these are helping to remove the inefficiencies and market distortions that lie behind many environmental problems.

Bioenvironmentalists and social greens challenge the logic and evidence behind these market-liberal arguments. Bioenvironmentalists see global trade (along with free-trade agreements) as a root cause of the global environmental crisis. The apparent wealth that seems to arise from free international trade in fact arises from the shortsighted exploitation of nature. Free trade externalizes environmental and social costs—that is, the prices of traded goods do not reflect their full environmental and social value (e.g., the price of a hardwood board does not reflect the value of a tree as a home for birds or a sink for carbon). Often, the price of products ignores, too, the environmental cost of the production process—for example, water and air pollution from a factory. In addition, *global* trade *distances* the point of manufacturing from the point of consumption, which leaves consumers less able or willing to accept the environmental consequences of their personal consumption habits. Like bioenvironmentalists, social greens also see trade as a root cause of global environmental problems. They agree with the points raised by bioenvironmentalists, but also stress the ways that trade exploits peoples and locales in unfair and unequal ways, resulting in uneven distribution of environmental and social problems. In particular, social greens argue that the global trading system contributes to the exploitation of developing countries (with respect to both workers and local environments).

Institutionalists take somewhat of a middle ground between these two broad positions on trade. On the one hand, they agree with the broad market-liberal view that trade can enhance prosperity, generate efficiencies, and result in higher overall capacity to manage environmental affairs. Yet institutionalists add that it is critical to manage trade judiciously. They argue that there are some cases—for example, trade in endangered species, hazardous wastes, and dangerous chemicals—where governments can and should control trade in order to protect the environment. For institutionalists, then, free-trade agreements need environmental clauses, and in some cases global environmental regimes need to override trade agreements. The key for institutionalists is for states to coordinate and cooperate on these issues. Following a brief overview of some basic data on globalization and trade, this chapter examines these three schools of thought on trade and the environment. To explore these arguments further in the context of real-world trade issues, the chapter then considers specific trade agreements, including the World Trade Organization as a global example, as well as several regional trade agreements.

Globalization and Trade

The process of economic globalization over the last 50 years, as we explained in chapter 2, has contributed to a sharp rise in the value and volume of global exports of goods and services (see figures 5.1 and 5.2). World trade as a percentage of gross domestic product (GDP) has also steadily grown from 25 percent in 1960 to 58 percent in 2001 (see figure 5.3). This high proportion of global GDP that is accounted for by trade indicates the growing importance of trade in the world economy. Indeed, from 1950 to 2001 global exports of goods rose twenty-fold.[2] The value of world merchandise exports was over US$6 trillion in 2002, dwarfing the value in 1948 (US$58 billion).[3] Much of this trade is between developed states, which accounted for three-quarters of world merchandise trade in 1996.[4]

While trade has grown remarkably on a global scale, there are still wide differences between and within regions (see figure 5.4). Western Europe (41 percent), Asia (23 percent), and North America (20 percent) together account for 84 percent of global trade in merchandise.[5] Latin America accounts for just 6 percent of global merchandise trade, Central and Eastern Europe and CIS states together just 5 percent, the Middle East just 3 percent, and Africa a mere 2 percent. There are equally wide

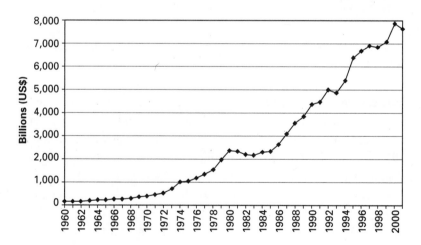

Figure 5.1
World exports of goods and services. *Data source:* World Bank World Development Indicators: www.worldbank.org/data.

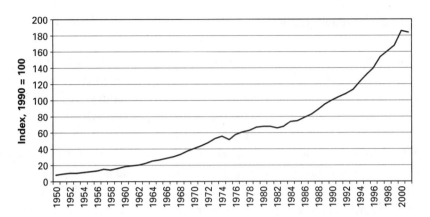

Figure 5.2
Growth in the volume of world merchandise exports. *Data source:* World Trade Organization Trade Statistics: www.wto.org/english/res-e/statis-e/statis-e.htm. WTO, 2003.

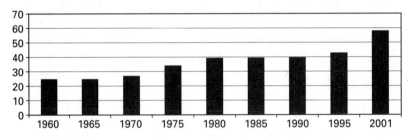

Figure 5.3
World trade (percent of GDP). *Data source:* World Bank World Development Indicators: www.worldbank.org/data.

differences over the value of trade for national economies within regions. In Africa, for instance, Rwanda's trade in goods and services accounts for 6 percent of its GDP, whereas in Equatorial Guinea trade accounts for 97 percent of GDP.[6] As the example of Equatorial Guinea shows, although Africa's proportion of total global trade is small, some economies rely heavily on trade and are thus quite vulnerable to market fluctuations.

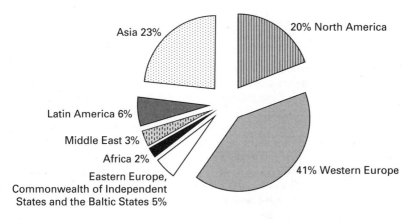

Figure 5.4
Percentage of global trade, by region (imports plus exports). *Data source:* World Trade Organization Trade Statistics: www.wto.org/english/res-e/statis-e/statis-e.htm. WTO, 2003.

Trade's Impact on the Environment: Three Schools of Thought

Trade Is Ultimately Good for the Environment
Most mainstream economists assume that in the long run the globalization of trade will enhance, rather than hinder, global environmental management. A few, like Columbia University professor Jagdish Bhagwati, have actively challenged the environmental critics of global trade.[7] These supporters, at a general level, accept the neoliberal economic assumption that trade promotes economic growth, and that economic growth in turn is crucial for preserving and managing the natural environment (for details, see chapter 4). The promotion of trade for market liberals is closely tied to this growth argument. But there are also more specific arguments as to why trade and environmental goals are mutually supportive. Trade, according to market liberals, enhances efficiency and diffuses cleaner technologies and standards. Each of these, in turn, promotes better environmental conditions. The efficiency from trade is based on the economic theory of comparative advantage (see box 5.1). This theory states that global welfare will rise (because of efficiency gains) if countries specialize in the products that they are relatively better at producing compared to other products, and then engage in trade with one another.

Box 5.1
Absolute and comparative advantage

The neoclassical economic assumption of the logical benefits of trade is grounded in two theories of advantage: absolute and comparative.

Absolute advantage suggests that if country A is able to produce a good (e.g., wheat) more cheaply than country B can produce it (perhaps for ecological, labor, technological, or societal reasons), and if country B is able to produce a different good (e.g., cloth) at a lower cost than country A can produce it, it is in the financial and material interests of *both* countries to specialize in the good they have the absolute advantage in producing and then trade these goods with each other.[1] In this case, then, both countries will have more of both goods (wheat and cloth) if they specialize their production in the product they have the absolute advantage in and engage in trade.

The theory of *comparative advantage*, first formalized in numerical terms by David Ricardo in 1817,[2] is consistent with absolute advantage, but applies another layer of logic that expands the instances when trade is beneficial. Imagine that country A has no absolute production advantage over country B in any product. Country B then produces both cloth and wheat more cheaply than country A can produce those goods. Under the theory of absolute advantage it does not seem logical for the countries to trade. Yet, as Ricardo shows, by thinking in terms of *comparative* advantage, trade is still logical, and indeed enhances the welfare of both countries. Ricardo (and many others since) demonstrated that if countries specialize in the goods that they themselves are *comparatively* best at producing (i.e., the products that they are relatively more efficient at producing compared to other goods), there will still be gains from trade.

To continue our example, if country A specializes in the good that it is relatively more efficient at producing (e.g., it is relatively better at producing wheat than it is at producing cloth), and then country A trades with country B for the product that country A is relatively less efficient at producing (in this case cloth), *total* productive output will still rise. This holds even though country B is absolutely more efficient than country A at making both cloth and wheat. This specialization and trade will create efficiencies, which in turn will increase the output of both products. And the competition among traders with similar or identical comparative advantages may foster further innovation and efficiencies.

1. Adam Smith (1776) outlines some of the most influential logic of trade.
2. Ricardo 1817. John Stuart Mill's 1848 work *Principles of Political Economy* ensured that comparative advantage would become central to the field of international political economy.

Though the theory of comparative advantage was not originally tied to global environmental management, today market liberals regularly note the environmental benefits of the efficiency gains from specialization. More efficient production optimizes the use of resources in the production process, meaning the world gets more goods using fewer resources.[8] More efficiency means more production per unit of inputs, which in theory should improve economic growth and raise per capita incomes. Higher incomes, market liberals argue, lead citizens to demand a cleaner environment (see chapter 4). More resources mean that more can be spent to preserve the environment as well as to enforce environmental regulations. Market liberals do not deny that rising production may initially harm the environment. The specialization needed to pursue comparative advantage may in fact result in some countries specializing in dirty industries. And if production rises faster than efficiency gains, pollution may also rise,[9] although often the increase in societal demand for a cleaner environment from higher incomes counteracts this. The key issue for market liberals is how the money is spent. As the WTO notes in a report on the linkages between trade and the environment, "The income gain associated with trade could in principle pay for the necessary abatement costs and still leave an economic surplus."[10]

More global trade, contend market liberals, generates other efficiencies beneficial to the environment as well. Trade policies that distort markets (such as tariffs, quotas, and export subsidies) can create situations where natural resources do not reflect real scarcity, which in turn drives overuse and overconsumption. Trade liberalization aims to remove these distortions and is thus needed to correct the underpricing of resources. Liberalization also diffuses more environmentally friendly production processes, because firms that produce goods behind trade barriers that protect them from competition have less incentive to adopt cleaner technologies.[11] The World Bank sums up this position well: "Liberalized trade fosters greater efficiency and higher productivity and may actually reduce pollution by encouraging the growth of less-polluting industries and the adoption and diffusion of cleaner technologies."[12]

Market liberals argue further that countries with more liberal trade policies are more likely to adopt higher environmental standards so as to better compete in markets with higher standards. The U.S. General Accounting Office found, for example, that U.S. standards for pesticide regulations are often taken into account by countries seeking to export agricultural products to the United States.[13] Another example is Mexico since NAFTA was signed. Evidence shows that environmental regulations in Mexico are now more stringent.[14]

For these reasons market liberals see the use of trade tools—such as sanctions (restrictions on trade to pursue other objectives)—to advance environmental goals as counterproductive. The fear is that such tactics could encourage countries to erect trade barriers under the guise of environmental protection. Bhagwati gives the example of the Ontario–United States "beer war" over a 10-cent-a-can tax on beer, which Ontario claimed was intended to discourage littering. The United States argued that the tax was discriminatory, because it was not levied on juice and soup cans and was borne disproportionately by the United States, since U.S. exporters supplied most of the canned beer for Ontario. The result was not a better environment in Ontario, but straightforward trade protection for Canadian producers of bottled beer.[15]

Partly to avoid discriminatory protectionism, market liberals are skeptical of the use of a precautionary principle that would enable countries to restrict trade in the absence of full scientific information on the grounds that an activity *may* harm the environment. Market liberals, for example, oppose the European Union's 1998–2004 moratorium on the import of genetically modified organisms (GMOs), seeing the use of the precautionary principle in this case as little more than governments protecting domestic markets. Interfering with trade, for market liberals, is rarely, if ever, the best tool to address environmental problems. "Trade sanctions are at best crude weapons," trade specialists Michael Weinstein and Steve Charnovitz argue, "and environmentalists should reconsider their enthusiasm for them."[16] For market liberals what is needed is sound environmental policy that zeros in on the source of the problem: indirect approaches through trade are inevitably inefficient and can even worsen environmental problems. Government bans on raw-log exports (to force exporters to process logs at home), for example, have made forest management worse in many countries. In Indonesia such a ban in the 1980s and 1990s allowed firms to export huge quantities of sawnwood and plywood, which wasteful and inefficient domestic firms processed with little regard for sustainable management.[17]

Certain market liberals go so far as to say that some countries have a higher capacity to absorb pollution than others, and that this is part of their natural comparative advantage.[18] The WTO notes that the existence of differing ecological conditions means that different national standards are bound to develop. It further notes that "even if ecological conditions were identical, international variations in standards may be desirable in order to reflect differences in income and ability to pay for environmental quality."[19] For market liberals, the use of trade sanctions to force a single, harmonized set of global environmental regulations would therefore destroy some of the natural advantages in some of the world's poorest regions. This would lower global welfare and ultimately hurt the natural environment.

In short, market liberals do not see free trade as a source of environmental problems. Instead, they argue that free trade brings efficiencies and growth. It reduces wasteful use of resources. It provides firms and governments with the funds to adopt or require environmental technologies. And it raises environmental standards in developing countries. Governments must address environmental problems with proper policies and incentives, and restricting trade does not create these policies and incentives. Critics of market liberalism, as we will see in the next section, have challenged all of these arguments, in many cases reaching the opposite conclusions.

Trade Is Bad for the Environment

Bioenvironmentalists and social greens are both skeptical of global free trade. In a nutshell, they associate many environmental problems with global free trade: the large scale and rate of economic growth under a free-trade regime; the unequal and unfair effects of the export-import synergies between the rich industrialized and poor developing countries; the hidden environmental and social costs of free trade; and the "dumbing down" of environmental standards.

Like market liberals, many bioenvironmentalists and social greens see global free trade as a potential catalyst for further global economic growth. But for bioenvironmentalists and social greens, the world, as we saw in chapters 2 through 4, cannot sustain such growth. A global-free-trade regime may well make production more "efficient," but such rapid and ever-upward growth will inevitably outstrip any gains in efficiency, so that the net impact on the natural environment is negative. This is referred to as the *scale effect*, where efficiency gains themselves can result in the use of even more resources. This can occur because more efficient production can lower consumer prices, which in turn can raise demand. The consumer in such cases may become thoughtless and wasteful (in effect thinking of the product as having little real value). The classic example is paper—supplies are plentiful and cheap. How many of us print stacks of papers we never read? How many of us would do that if each sheet of paper was worth a bottle of wine? Demand for a product may also simply outpace any potential efficiency gains. Cars, for example, are far more fuel efficient now than they were three decades ago, yet there are so many now that the total amount of resource use is much higher. The expansive nature of international trade is so closely tied to the growth of consumption that the ecological impact of the sheer increase in volumes far exceeds any potential efficiency gains that derive from more trade.[20]

Social greens in particular add that global "free" trade does not benefit all trading partners equally. The original theories of the benefits of trade under conditions of absolute and comparative advantage assume that goods, but not capital or labor, are mobile. This may have made sense when the theories were first postulated several hundred years ago. But this premise, social greens argue, is no longer valid. Capital is highly mobile in today's world, which fundamentally alters the relationship between trade and welfare, as well as trade and the environment.[21] Today, specialization for trade concentrates pollution in production zones— and usually it is developing countries that absorb most of the environmental costs of free trade by producing the most polluting items, while rich countries enjoy the benefits of those products. Moreover, global trade merely delays the inevitable limits to growth by distributing environmental costs worldwide, allowing some countries to consume far beyond the carrying capacity of their natural environment, at the same time using up the carrying capacity of other countries. "Free trade," ecological economist Herman Daly argues, "allows the ecological burden to be spread more evenly across the globe, thereby buying time before facing up to limits, but at the cost of eventually having to face the problem simultaneously and globally rather than sequentially and nationally."[22]

Production for export in the developing world tends to rely heavily either on the unsustainable use of local natural resources or on dirty and unsafe factories (relying on cheap labor). Plantations of coffee, groundnut, and palm oil (for cooking oils and margarine) in Latin America,

Africa, and the Asia-Pacific region show the environmental harm of export-oriented crops.[23] Forestry, fishing, and mining in these regions highlight the ecological damage of large-scale natural resource extraction for export.[24] Across the developing world, textile and electronics factories pollute the environment with mass production of cheap goods for consumers in wealthy countries.[25] Trade, in other words, allows countries to import or export carrying capacity, so the rich may live beyond their carrying capacity, while the poor live well under it. Some, as the previous chapter mentions, refer to this as part of a country's ecological shadow;[26] others call this "distancing."[27] Such trade in effect casts ecological harm from consumers onto countries that are net exporters of natural resources or pollution-intensive goods (or both). When the shadow cast is long and the distance is great (either in geographic or mental terms), there is little ecological feedback to the consuming countries.

There are other innate drawbacks of too much trade. The prices of goods and services in a free-trade regime often do not reflect the true environmental and social costs of production. This problem generally begins with the extraction of natural resources. The market value of an old-growth tree, for example, primarily reflects the cost of access to the forests (perhaps a tax or flat fee), along with the cost of harvesting and transportation to the mill. The tree (now a log) is sold to a mill for a "profit" when the buyer's price is higher than the total monetary costs for the logger. Yet little, if any, effort is made by governments to assign a monetary value to the hundreds of years it took for the tree to grow. Or to the aesthetic or ecological value of the tree in a forest. Or to the global environmental value of the tree as a store and absorber of carbon dioxide. Instead, the tree is worth what the government of the day sees as its financial value to the people of the nation-state. In many cases it is worth almost nothing. In fact, getting it out of the forest and into "production" for the international market creates jobs as well as income and spurs economic growth, and most governments merely assign a token fee for access. The stumpage value—or the total state fees for a tree in the forest—has been, for example, typically less than 1 percent of the consumer price in the Asia-Pacific region.[28] Accounting for the "true value" of timber would create a very different trading structure, one in which few, if any, old-growth trees would ever enter into world trade.[29]

The externalization of the environmental costs of traded goods does not end here, but continues throughout the production process (from extraction to the exporter to the importer to the factory floor to the retail outlet). The price a consumer pays, for example, generally does not include the full costs of transportation. This occurs partly because governments tend to subsidize energy inputs (in a bid to spur growth and industrialization), and partly because energy sources themselves (like oil and coal) generally do not reflect the full environmental costs (e.g., the climatic effects of the release of carbon dioxide into the atmosphere).[30] The underpricing of the transportation of goods over long distances—via ships, trucks, or airplanes—encourages more trade, and the end result is more fuel use, greenhouse gas emissions, and smog.[31] Moreover, even if properly priced, the transportation effects of trade will cause significant environmental disruptions.[32]

Critics further argue that trade puts downward pressure on environmental standards. This occurs because countries, in a bid to reduce costs to become more competitive in international markets, sometimes lower or at least fail to strengthen environmental regulations as well as enforcement.[33] Some refer to this as the "race to the bottom."[34] Others point out that many countries appear "stuck at the bottom."[35] Critics argue that weak regulations and enforcement in order to gain a competitive edge simply encourages firms to externalize environmental costs. Many critics further argue that the global environmental standards that trade agreements rely on may be less strict than the domestic standards of the original signatories, thus pushing down environmental regulations in advanced economies. For example, Codex Alimentarius—a UN commission that sets global food-safety standards for trade—adopted standards for pesticides in the 1990s that were 42 percent lower than the standards of the U.S. Environmental Protection Agency and the U.S. Food and Drug Administration. The Codex standards allowed farmers to spray fifty times more DDT on peaches and bananas than the U.S. standards allowed.[36]

Social greens in particular argue further that most global trade agreements, even ones with environmental clauses, prioritize economic concerns over ecological ones, partly because corporate interests—sometimes through firms and sometimes though states agencies—wield so much power over the agendas and negotiations.[37] A study by the activist group Public Citizen, for example, found that trade advisory panels in the United States in the early 1990s were dominated by industry players, many of which had either been fined for environmental infractions or had lobbied to defeat environmental laws within the United States.[38]

In view of the numerous problems associated with trade, many social greens and bioenvironmentalists argue that it is perfectly legitimate—indeed, it is often necessary—for states and global institutions to restrict trade to achieve environmental goals. Trade sanctions (or the threat of sanctions), for instance, can be an effective (and nonviolent) way to punish states that duck global environmental standards to "free ride" on the global benefits of higher environmental standards in other countries. These thinkers stress that the only way to achieve a sustainable economy is to limit international free trade and its associated growth.[39]

Moreover, both bioenvironmentalists and social greens support the use of the precautionary principle in trade agreements to prevent future environmental harm.[40] Waiting for full scientific consensus before taking action, they argue, will often, if not always, mean the action is too late to prevent serious (and perhaps irreparable) ecological damage. DDT is a good example. In 1948, Paul Müller of Switzerland was awarded the Nobel Prize in medicine and physiology for his 1939 discovery that DDT was a potent insecticide. Since then it has saved millions of lives, killing mosquitoes that spread malaria and lice that transfer typhus. Yet DDT, scientists learned in the decades after Müller's Nobel Prize, slowly builds up in the fatty tissues of animals (partly because it is chemically stable and is soluble in fat). It is also relatively toxic for fish. The dangerous ecological and health side effects led the United States to ban DDT in 1972. Today, over thirty countries ban or greatly restrict the use of this substance. Other countries, however, still use it as an insecticide for agriculture and disease management, and the international community is just beginning to develop global mechanisms to control the production and trade of DDT with the signing of the 2001 Stockholm Convention on Persistent Organic Pollutants (in force as of 2004; see chapter 3).

The Stockholm Convention does not ban DDT. Under the rules Parties can notify the Secretariat and then use DDT for disease control until other cost-effective and environmentally friendly chemicals can replace it. DDT, we now know with scientific certainty, damages the liver as well as the nervous and reproductive systems; it is also a probable cause of cancer. Yet long delays and political compromises, and the lack of funds for research on alternative means of disease control, have meant that

DDT still enters the atmosphere, and will continue to contaminate water systems and food chains for years to come. As a result, infants throughout the world will continue to consume DDT through breast milk. And children and adults throughout the world will continue to consume DDT through local fish and shellfish as well as food imports from countries contaminated with (or still using) DDT.[41]

Managed Trade Can Be Good for the Environment

Arguments over whether trade is good or bad for the environment dominated the initial debate over trade and the environment throughout most of the early 1990s. A compromise position, along the lines of institutionalist thought, garnered increasing support after the Rio Earth Summit in 1992. The Summit went to great pains to emphasize that free trade and environmental protection were not in conflict. Principle 12 of the Rio Declaration, for example, says that "states should cooperate to promote a supportive and open international economic system" and that "trade policy measures for environmental purposes should not constitute a means of arbitrary or unjustifiable discrimination or disguised restriction on international trade." Further, Agenda 21 states that "the international economy should provide a supportive international climate for achieving environment and development goals by . . . promoting sustainable development through trade liberalization."[42]

Advocates of "managed trade" accept the broad claims of market liberals with respect to the environment.[43] But they qualify those claims. They argue that international trade need not be any more harmful to the environment than regional or local trade. Each kind of trade involves a mixture of benefits and problems. States are no doubt tempted to weaken environmental standards in order to gain a competitive advantage in trade, yet global agreements can largely resolve this drawback of free trade.[44] Global cooperation, in other words, is an effective mediating force that can produce positive outcomes for both trade and the environment.[45] Such cooperation is in fact essential because efficient trade requires healthy environments, while healthy environments (at least in an era of globalization) will degrade and collapse without trade (a glance at the environment of North Korea confirms this).

Sometimes, institutionalists argue, global agreements warrant trade measures, on occasion even trade sanctions. Obvious cases include the need to control the global trade in hazardous wastes and dangerous chemicals (such as persistent organic pollutants, ozone-depleting substances, and banned chemical pesticides). Yet the global community should only rely on such measures in clear instances of environmental harm resulting from trade. At the same time, institutionalists argue that it is important to liberalize trade to improve global environmental management. The elimination of subsidies on pesticides and fuels is an example of how liberalization of trade can help promote cleaner environments.[46]

Furthermore, global cooperation can create green markets—that is, markets where the prices throughout a trade chain better reflect the full environmental and social costs. This can lower consumption and increase the amount of funds available for sustainable management. Global ecolabels for products from "sustainable sources" or with "sustainable production processes," for example, can encourage consumers (as well as perhaps wholesalers and retailers) to voluntarily pay more for a product that does less damage to the global environment. The market growth of timber certified as sustainable and coffee labeled as organic or fair trade suggest the potential here for win-win trading arrangements. The nonprofit Forest Stewardship Council (FSC), founded in 1993 to promote better forest management, is an example of an organization designed to construct a green market. Its members include environmental NGOs, forest industries, indigenous and community groups, and timber-certification bodies. The FSC accredits and monitors national certifying agencies. Its logo on a wood product is meant to signal to the consumer that "the product comes from a well-managed forest."[47]

Institutionalists who back managed trade also tend to advocate a limited use of the precautionary principle. While they do see the potential for those states wishing to protect domestic industries to abuse this principle, they also believe there are some instances where precaution is warranted.[48] They also support the "polluter-pays principle" (PPP). This principle states that the polluter should bear the full cost of pollution prevention and control, "without public subsidy, as means of encouraging the internalisation in the price of goods of the full costs of pollution abatement."[49] Often, government taxation schemes as well as subsidies to promote domestic industry relieve consumers of the burden of paying for the costs of cleanup, or for the long-term harm to the global atmosphere by production at, say, a smokestack factory. In such cases, governments may need to consider restrictions (such as tariffs or even bans) on the import of polluting goods, or on goods produced with considerable environmental harm.[50] Firms and governments could then take the profits from such tariffs and cost internalization and invest them in sustainable production methods—or perhaps, more radically, use these funds to compensate countries for past environmental degradation.[51]

Proponents of managed trade see the greening of international trade agreements as one of the most constructive and practical ways to move toward both a richer and a cleaner world. And they argue that such agreements should coordinate with international environmental agreements in ways that produce win-win situations.

Greening of International Trade Agreements?

Theoretical differences over the environmental merits and demerits of trade continue to influence international debates and negotiations. At the same time, however, much of this debate now takes place in the context of real-world trading arrangements, many of which now explicitly address environmental management. This section explores these debates through a detailed case study of the WTO as well as discussion of several regional trade agreements, including NAFTA, the European Union (EU), and the Asia Pacific Economic Cooperation (APEC).

The WTO: Principles of Trade

The World Trade Organization, founded in 1995, evolved out of the General Agreement on Tariffs and Trade (GATT). The GATT was established in 1947 as a global trade agreement that aimed to achieve freer trade. It was founded on the neoclassical economic assumption that free trade will improve global welfare, following on the ideas of David Ricardo (see box 5.1). The principles of the GATT are still in force today under the rubric of the WTO. One of the key principles is the most-favored-nation (MFN) principle, which requires each GATT signatory to accord to all other GATT signatories the same trading conditions as it accords to its "most favored nation." Signatories must also provide so-called national treatment to other GATT signatories. This requires them to treat "like products" the same regardless of the country of origin or the process of production. The GATT also calls for a general reduction in the level of tariffs and an elimination of quantitative restrictions on trade. The GATT secretariat was an interim body to oversee the trade agreement.

The GATT was periodically renegotiated through eight trade "rounds." The 1986 Uruguay Round, completed in late 1994, created the WTO as a permanent body to oversee global trade rules. The WTO adopted the rules negotiated under the GATT (which deals with trade in goods), as well as a number of other agreements, including the Trade Related Intellectual Property Rights Agreement (TRIPS Agreement), the General Agreement on Trade in Services (GATS Agreement), the Agreement on Technical Barriers to Trade (TBT Agreement), the Agreement on the Application of Sanitary and Phytosanitary Measures (SPS Agreement), and the Agreement on Agriculture (AoA), which were all negotiated as part of the Uruguay Round. The WTO also created a Dispute Settlement Body and adopted a dispute-resolution process that is compulsory and binding, properties that the GATT dispute-resolution process lacked.[52] A permanent Appellate Body to hear appeals to WTO dispute-panel decisions was also created. Countries found in violation of the WTO rules do not have to remove the offensive trade restriction if they are willing to compensate complainants by allowing them to apply retaliatory trade sanctions.

Following the failed attempt to launch a new round of trade negotiations under the WTO in Seattle in late 1999 (see chapter 3), a new round was finally launched in 2001 in Doha, Qatar (referred to as the Doha Round).[53] Talks are ongoing, though negotiations at the WTO ministerial meeting held at Cancún, Mexico, in September 2003 collapsed due to disagreements between industrial and developing countries over a number of issues, including most prominently revisions to the Agreement on Agriculture. Today 146 countries are members of the WTO, and GATT/WTO trade rules now cover over 90 percent of world trade.[54] WTO members pledge to obey the provisions of the WTO agreements. If one country believes another country has violated these provisions (resulting in a negative impact on the first country), it turns to the dispute panel for a ruling.

The GATT was created well before major concerns arose over the state of the global environment. It therefore contains few direct references to the environment. The preamble to the WTO does mention the need to pursue sustainable development, making it somewhat more progressive than the original GATT agreement. But the GATT is still the main trade agreement for goods under the WTO. The main article in the GATT that appears to potentially accommodate environmental concerns is Article XX (see box 5.2). This article sets out cases eligible for exceptions to the GATT rules. It is not terribly clear, however, on whether exceptions are allowed for environmental reasons.

Article XX allows exceptions to the GATT rules to protect human, animal, or plant life or health, or to ensure conservation of natural resources. Yet even these exceptions are qualified. If a country uses trade restrictions to protect the health of humans, animals, or plants under XX(b), such measures must be considered *necessary*. That is, if other measures that do not restrict trade are available, then trade restrictions are not seen as necessary. If a country restricts trade to conserve exhaustible natural resources under XX(g), such measures must *relate strictly to natural resource depletion, and domestic restrictions must also be instituted at the same time*. That is, if a country restricts trade, it *must* simultaneously impose domestic restrictions.[55] Moreover, any trade measures under this article *cannot be arbitrary or unjustifiable*, as stated in the introductory paragraph (referred to as the chapeau).

More broadly, Article XX does not cover global environmental concerns very well. Many environmental problems do not fit into its narrow terms, such as global warming or dumping at sea, and thus do not qualify for exceptions to GATT rules. Measures seeking to protect natural

Box 5.2
GATT article XX (general exceptions clause of the GATT)

Subject to the requirement that such measures are not applied in a manner which would constitute a means of arbitrary or unjustifiable discrimination between countries where the same conditions prevail, or a disguised restriction on international trade, nothing in this Agreement shall be construed to prevent the adoption of enforcement by any contracting part of measures: . . .

b) necessary to protect human, animal or plant life or health; . . .

g) relating to the conservation of exhaustible natural resources if such measures are made effective in conjunction with restrictions on domestic production or consumption.

resources outside of the restricting country's border also are not allowed—that is, the trade measures cannot be extrajurisdictional or uni-lateral, and can only apply to natural resource depletion or animal, plant, or human health in the country imposing the measure. That said, more recent dispute-panel decisions have broadened the scope of WTO excep-tions somewhat (as discussed later).

Furthermore, the national-treatment rules of the WTO stipulate that countries cannot restrict trade based on the country of origin or on *how* goods are produced. This in effect means that countries cannot discrim-inate against products based on production and processing methods (PPMs) if the products are otherwise identical. Article XX only reinforces this. Countries can only restrict trade based on *what* is produced—that is, on the material composition of the products. This occurs because WTO trade rules call for no discrimination between "like products," where like products are those that are similar in their physical qualities, regardless of whether the method of production was clean or dirty. But these rules do allow countries to impose restrictions because of physical characteristics of products, such as the chemical content, product emis-sions levels, and the like.[56]

Article XX and Environment Disputes at the GATT/WTO

Given the potential ambiguity of GATT/WTO rules for environmental protection, and given the stringent conditions that must be met, it is not surprising that to date, there have been a number of environment-related trade disputes at the GATT/WTO (see table 5.1). Some countries have cited Article XX as a rationale to justify trade restrictions for environ-mental reasons, but these have nearly all been ruled as GATT-illegal. The Tuna-Dolphin cases of 1991 and 1994 are two of the best known of the trade-environment disputes at the GATT/WTO. In the 1991 dispute, Mexico complained that the U.S. Marine Mammal Protection Act (MMPA) called for an embargo against imported tuna from countries that could not prove that their fishing methods did not harm dolphins. The GATT panel decided against the United States, and in favor of Mexico. The United States tried to justify its action based on GATT Article XX(b), by arguing that its actions were "necessary" to protect human or animal life; it also relied on Article XX(g), by claiming that it undertook the measures to conserve natural resources. Specifically, the

Table 5.1
Key GATT/WTO environmental dispute-resolution panels

Date filed	Government(s) requesting panel	Government being challenged	Relevant provision	Issue	Status
Sept. 3, 1991 (report date)	Mexico	U.S.	Goods: general exceptions of GATT Article XX	U.S. placed embargo on imports of tuna that did not meet dolphin-protection standards set out in U.S. law. U.S. also required tuna products to be labeled "dolphin-safe."	GATT panel ruled against U.S. embargo. Labeling was permitted if targeting domestic and foreign producers. Report circulated but not adopted. Mexico and the U.S. held bilateral consultations aimed at agreement outside GATT.
June 16, 1994 (report date)	European Union and Netherlands	U.S.	Goods: general exceptions of GATT Article XX	U.S. extended embargo (above) to "intermediary" countries handling the tuna en route to the U.S.	1994 panel report upheld some findings of the first "tuna" panel and modified others. U.S. claimed insufficient time to complete its studies; therefore no consensus to adopt the report, a requirement under the old GATT system.

Jan. 23, 1995	Venezuela	U.S.	Goods: general exceptions of GATT Article XX	1990 amendment to the U.S. Clean Air Act applied stricter rules on the chemical characteristics of imported gasoline than it did for domestically refined gasoline.	WTO panel and Appellate Body decided that U.S. rules were "discriminatory."
Jan. 31, 1996	U.S.	European Union	Agreement on the Application of Sanitary and Phytosanitary Measures (SPS Agreement)	EU ban on hormone-treated beef (applies equally to European farmers and foreign producers).	A WTO panel ruled against the EU ban.
Oct. 14, 1996	India, Malaysia, Pakistan, Philippines, and Thailand	U.S.	Goods: general exceptions of GATT Article XX	U.S. ban on shrimp caught in nets without "turtle-excluding devices" (applies equally to domestic and foreign producers).	WTO panel found U.S. regulations in violation of trade rules. An appeals panel upheld the decision, but on narrower grounds that the U.S. regulations had been applied in a discriminatory fashion.

Table 5.1
(continued)

Date filed	Government(s) requesting panel	Government being challenged	Relevant provision	Issue	Status
June 3, 1998	Canada	France	Agreement on Technical Barriers to Trade (TBT Agreement)	French embargo on imports of asbestos and asbestos-containing products.	Panel and Appellate Body rejected Canada's challenge (WTO agreements support members' ability to protect human health and safety at the level of protection they deem appropriate).
May 20, 2003	U.S., Canada, and Argentina	European Union	Agreement on the Application of Sanitary and Phytosanitary Measures (SPS) Agreement	EU moratorium on approval of genetically modified foods and seeds.	Pending.

Source: WTO website: www.wto.org/english/tratop_e/envir_e/edis00_e.htm.

United States argued that the regulation of non-product-related PPMs—in this case tuna fishing that was harmful to dolphins—outside of its own territory was necessary to protect dolphins. The GATT panel did not fully accept this argument, and ruled that Article XX was not applicable in this case. The U.S. actions were seen as unilateral and extrajurisdictional (i.e., application of national laws outside the United States) such that they discriminated against "like products," thus violating the national-treatment clause of GATT.[57] Moreover, the panel pointed out that the United States could have pursued less trade-restrictive, multilateral efforts to reduce dolphin kill. It did say dolphin-safe labels were allowed, however.[58]

Various interpretations of the ruling have been put forward. Some have argued that the way the MMPA operated was flawed and thus went against GATT rules. In particular, it determined permissible dolphin kill based on what U.S. fishermen had achieved that same season, something Mexican fishermen could not know in advance. Yet many were critical of the GATT ruling against "extraterritoriality," which was viewed as artificial and inappropriate to contemporary cross-border environmental problems.[59] In the second Tuna-Dolphin dispute of 1994 the European Union challenged the secondary embargo imposed by the United States on tuna from third-party sellers that did not meet U.S. regulations. The GATT dispute panel again ruled against the United States, claiming that the application of the MMPA in this case was unilateral and arbitrary, and thus did not qualify for an exemption under Article XX.

The WTO has ruled on other environment-related disputes as well. In 1996 Brazil and Venezuela complained to the WTO that U.S. laws on gasoline cleanliness discriminated against gasoline imports from these countries. In this case, a 1990 amendment to the U.S. Clean Air Act required refiners to clean up fuel relative to 1990 baseline standards, starting in 1995. The law stipulated that domestic refineries could use their own baseline from 1990, but that foreign refineries had to clean up their fuel exports to the United States relative to average 1990 U.S. gasoline quality. Brazil and Venezuela won this dispute, and as a result the U.S. Environmental Protection Agency had to revise its laws on gasoline cleanliness. The WTO panel ruled that, although in principle Article XX(g) covered the standards of the Clean Air Act since these aimed to protect exhaustible resources, the way the United States set up and applied these baselines constituted unjustifiable discrimination. Specifically, it said that the United States could have been less discriminatory to foreign refiners by applying the same rules for baselines for both domestic and foreign firms.[60] Again, environmental activists interpreted this ruling as being "antienvironmental." Others argued, however, that the U.S. law was plainly discriminatory, and that the WTO did not rule against environmental protection but rather against a discriminatory law.

Environmental groups also saw the 1996 Shrimp–Sea Turtle case as proof the approach of the WTO was antienvironmental. In this case, Thailand, Malaysia, Pakistan, the Philippines, and India all complained about a U.S. law discriminating against imports of shrimp that did not use turtle-excluding devices in their nets, which U.S. law requires. These countries claimed they were taking other measures to protect turtles. The United States tried to claim an exemption to WTO rules under Article XX. The WTO dispute panel ruled, again, that the U.S. requirement that other countries observe its domestic laws was not legal under WTO rules. This behavior was deemed to be discriminatory toward a product based on its production methods, and ineligible for exemption under Article XX because it was a unilateral measure. The United States appealed the case, and the Appellate Body altered the ruling somewhat. It ruled that in principle Article XX could cover the U.S. measures, but it added that these were nevertheless applied in an unjustified and arbitrary manner because it assumed that all countries could meet the requirements of the U.S. law equally.[61] The Appellate Body also ruled that the United States did not make enough effort to negotiate a sea-turtle conservation agreement with the complaining countries. The WTO prefers multilateral efforts to deal with such problems, not unilateral measures.

These rulings appear to put trade ahead of the environment. But some argue that a closer inspection shows that the WTO is in fact accommodating many environmental concerns.[62] In the case of the Shrimp–Sea Turtle-ruling, the decision that the U.S. measures in principle could be taken under Article XX(g), provided they were not applied in an unjustified and arbitrary manner, was very different from the Tuna-Dolphin ruling, which held that the measures to protect dolphins were not

allowed at all under Article XX because they applied outside the jurisdiction of the United States. This ruling also indicates that the WTO is
somewhat tolerant of the idea of discrimination based on production and
processing methods.[63]

Other WTO Agreements and the Environment

A number of the other agreements under the WTO from the Uruguay
Round also mention the environment in one way or another.[64] The Agreement on the Application of Sanitary and Phytosanitary Measures (SPS
Agreement) relates to food safety and human, animal, and plant health
and safety regulations. The purpose of this agreement is to ensure that
states do not adopt safety measures that create unfair obstacles to trade.
The agreement includes exceptions that allow a country to adopt an SPS
measure in order to protect the environment, as well block trade for a
limited period of time in the case of scientific uncertainty. In these cases,
measures must be subjected to risk assessment and scientific verification.
This has been interpreted as a very limited use of the precautionary principle at the WTO. The WTO encourages countries to harmonize their
SPS standards or to adopt internationally recognized standards, such as
those developed by the Codex Alimentarius Commission of the Food and
Agriculture Organization and World Health Organization. However,
countries are allowed to have more stringent measures than those set by
the Codex. SPS measures also must meet other criteria, in particular that
they not be applied in an arbitrary or unjustified way.

Several disputes have been brought forward to the WTO claiming violation of the SPS Agreement. In 1996 the United States and Canada, for
example, challenged the ban by the European Union on imports of beef
treated with hormones. The WTO dispute panel ruled against the
Europeans on the grounds that the ban was not based on a valid risk
assessment.[65] Though the European Union lost this case, it maintained
the ban and faced the trade sanctions the WTO allows as retaliation for
violating the agreement. The European Union's 1998 moratorium on the
import of GMOs was undertaken on the grounds that the moratorium
was justified under the SPS Agreement, with the European Union declaring that the ban on the import of such products was temporary while
it conducted scientific assessments. The United States, Canada, and
Argentina initiated a challenge to the European Union at the WTO in

May 2003.[66] This prompted the European Union to pass tough new laws on the labeling of GMOs in July 2003. This enabled the European Union to lift its moratorium on the import of GMOs in 2004. It is not clear, however, that the complainants will drop the challenge, because they are arguing that the new laws are equally unjustified under the SPS Agreement.

The Agreement on Technical Barriers to Trade (TBT Agreement), which seeks to ensure that countries do not use technical regulations and standards to block trade, also allows for countries to adopt technical regulations to protect human, animal, or plant life or health, or the environment. Again, these rules are subject to qualifications. In this case, such measures must be transparent and nondiscriminatory. Like the SPS Agreement, the TBT Agreement encourages countries to harmonize standards or to adopt internationally set standards where they exist, but again allows for exceptions where necessary. One dispute has been brought to the WTO that relates to the TBT Agreement and the environment. This is the Canadian challenge against the European Union regarding a French law banning the import of asbestos and asbestos products. Initially, the dispute panel ruled that the ban was not a technical regulation under the TBT Agreement and that it was allowed under Article XX(b) of the GATT. Following an appeal by Canada, however, the Appellate Body ruled that it was indeed a technical regulation, but nevertheless it upheld the ruling that the ban was allowed under Article XX(b).[67]

The General Agreement on Trade in Services (GATS Agreement) also incorporates an article similar to Article XX of the GATT, which allows for exemptions if necessary to protect human, animal, or plant life or health, but not relating to the conservation of exhaustible natural resources. But again, such measures must not be arbitrary or unjustifiable. The Trade Related Intellectual Property Rights Agreement (TRIPS Agreement)—which sets rules on intellectual property, such as patents and copyright, and on trade—also has significant environmental implications. It calls on countries to make patents available for inventions—involving both products and processes. This includes microorganisms and biological processes for the production of plants or animals. Plant varieties should be given protection either by patents or by an effective sui generis (designed for a specific purpose) system. But it allows WTO members to exclude from patentability any inventions that may endanger the environment, provided it can be shown that doing so is necessary for environmental protection. Finally, the Agreement on Agriculture (AoA), which seeks to reduce subsidies on agriculture, allows countries to continue to subsidize agriculture in ways that do not distort trade (such subsidies are referred to as the "green box"). Included in the subsidies allowed under the agreement are those that relate to protection of the environment, though these must be part of clearly defined government environmental or conservation programs. No environmentally related challenges have yet been brought forward regarding the AoA, TRIPS, or GATS.

The Committee on Trade and Environment and Potential MEA-Trade Conflicts

The WTO has made other explicit efforts to improve its image on the environmental front over the past decade. In 1994 the General Council of the WTO reconstituted the Committee on Trade and Environment (CTE) (a group that was set up in the 1970s but that had not met in years) as a permanent body to discuss the relationship of trade and the environment. This committee, which now meets periodically, discusses how the WTO should handle issues such as the link between trade agreements and Multilateral Environmental Agreements (MEAs), ecolabeling, and the ways in which trade and the environment relate to trade rules regarding intellectual property and services. The CTE is charged with making recommendations on these items, including whether to amend the WTO agreements to more explicitly spell out rules on cases where environmental goals can supercede trade goals.

So far there has been a great deal of talk in the CTE, but not much concrete action on these fronts. The slow progress is partly due to opposition from developing countries, and partly due to a lack of leadership on the part of the United States.[68] Developing countries are especially worried about ecolabeling rules now under discussion, because they fear these will allow developed countries to discriminate against them for not labeling products with environmental information about the production process. Such labeling schemes are difficult and expensive for poor countries to implement. The issue of trade and the environment did not even appear on the official agenda at the failed launch of trade talks in Seattle in November-December 1999. It did, however, make it onto the agenda at the WTO ministerial meeting in Doha in 2001, which launched the Doha Round of trade negotiations. Ministers agreed at that meeting to undertake negotiations for the purpose of clarifying the relationship between MEAs and WTO rules.[69]

The question of the relationship between MEAs and WTO rules warrants further elaboration here, because this is an area where there is potential for a trade dispute and the question has sparked considerable debate. There are some 200 MEAs, and about 20 of them include trade provisions. In other words, the environmental agreements restrict trade in some way or other as a means to meet an environmental goal. Some of the more prominent MEAs with trade provisions include the Basel Convention, the Convention on International Trade in Endangered Species, the Montreal Protocol, the Kyoto Protocol, the Cartagena Protocol, the Stockholm Convention, and the Rotterdam Convention (for details, see chapter 3, table 3.1). In some cases, an MEA might restrict trade in a risky or endangered product that it is seeking to control. In others cases, it might restrict trade between parties and nonparties in certain products as a means of encouraging countries to join the agreement, and because it would prevent nonparties from exploiting markets for products the agreement is seeking to regulate.[70] Many MEAs include other control measures as well that may affect trade, such as provisions for technology transfers and prior informed consent.

MEAs use such measures in an effort to ensure compliance and meet environmental goals. But some analysts argue that these have the potential to be WTO inconsistent. That is, they could be interpreted as being incompatible with WTO rules on most-favored-nation and national-treatment issues, as well as with WTO requirements to eliminate quantitative restrictions. Such inconsistencies are seen to be most problematic when one country that is party to a multilateral environmental agreement applies trade restrictions against a country not party to the MEA, yet both countries are WTO members. An example is the Montreal Protocol, which imposes more stringent restrictions on trade in ozone-depleting substances, or products produced using them, between a party and a nonparty than between two parties. In cases where both countries are party to an MEA (and thus have voluntarily endorsed the provisions of the agreement), most analysts believe it is unlikely that a conflict over trade and environment rules would reach the WTO. But problems could arise in these cases, too. The Basel Convention, for example, bans the trade in hazardous waste between parties listed in the convention's annex VII (mainly non-OECD countries) and those that are not. Some parties to the convention could, in theory, argue that this is discriminatory because it discriminates based on a country's membership in the OECD. Other types of MEA trade-related measures, such as those that deal with the precautionary principle and intellectual property rights, also have the potential to create tensions.[71]

Thus far, no cases have come forward that officially challenge MEAs as being WTO inconsistent. But environmental activists are concerned about the growing number of environmental agreements that certain countries have refused to ratify. For example, certain key environmental agreements, including the Basel Convention, the Cartagena Protocol, and the Kyoto Protocol, still lack ratification from a number of countries, most notably the United States. In such cases where there are significant economic players who are not parties to MEAs with trade provisions, the potential for a challenge at the WTO is heightened. The outcome of any such dispute would be binding under the WTO rules, and would set a precedent for future cases.

There are several ways the WTO might address the potential inconsistencies between the WTO rules and MEA rules. A number of proposals have been discussed. One would be for the WTO to waive certain trade obligations when trade restrictions are imposed for MEA purposes. Another would be to amend the WTO rules to list MEAs exempted from trade challenges.[72] The WTO could also draw up rules that trade measures in MEAs would have to follow to be consistent with WTO rules.[73] Changing the WTO rules, however, can only be done by consensus, and not all countries agree on the need for change. Short of amending the WTO rules, the clarification of the relationship could by default be determined by dispute-panel resolutions.

The relationship between MEAs and the WTO was debated at the Johannesburg World Summit on Sustainable Development in 2002. The Plan of Implementation from the Summit calls on signatories to "promote mutual supportiveness between the multilateral trading system and the multilateral environmental agreements, consistent with sustainable development goals, in support of the work program agreed through the WTO, while recognizing the importance of maintaining the integrity of both sets of instruments."[74] However, the CTE talks have not yet yielded any concrete recommendations on this front.

Reforming the WTO?

While it could be argued that these steps taken within the WTO indicate that some progress is being made in that institution regarding environmental considerations, there still remains disagreement about whether and the extent to which such change should be taking place. Most market liberals argue that the WTO rules as currently constituted are sufficient for dealing with environmental concerns, and extreme market liberals argue that environmental issues have little or no place in trade policy. More moderate market liberals do acknowledge that some minor reforms to the WTO could help to spell out its relationship to the environment more explicitly. Most market liberals see the merits of clarifying the relationship between MEAs and WTO rules, for example, but some are still skeptical, and want to be sure that environmental agreements represent genuine environmental efforts and are not just protectionism in disguise. Increased transparency and clear rules for ecolabeling are also sought.[75]

Such changes to the WTO are deemed minor, and overall most market liberals argue that WTO rules are green enough. The liberalization of agricultural trade is a good example of this point. The aim of the WTO's Agreement on Agriculture is to reduce tariffs on agricultural products and to eliminate or reduce agricultural subsidies. Market liberals argue that such measures are key not just for facilitating trade and promoting growth, but also for environmental protection, because the removal of trade-distorting tariffs and subsidies will make land use much more efficient.[76] Working toward these original aims of trade liberalization, then, is the best policy for the environment according to these thinkers. Market liberals are also wary of incorporating a precautionary principle into the WTO's rules, which they see as an open invitation for trade protectionism.

Institutionalists have tended to argue for more substantial reforms to the WTO. They see the merits of a WTO, and with some substantial retuning see it as essential in the quest for sustainable development. Insti-

tutionalists agree with market liberals on the need for a clearer deline-
ation of the way MEAs and the WTO should interface, though they are
stronger advocates of the use of trade sanctions in MEAs for achieving
environmental goals. They also agree on the need for explicit rules for
ecolabeling, and on the necessity of transparency, accountability, and
participation from civil society in the WTO processes.[77] Going further
than market liberals, most institutionalists back the limited use of the
precautionary principle in dispute resolutions at the WTO and accept
discrimination based on product process or methods.[78] Some go so far
as to argue for a new organization to counter the power of the WTO,
in the form of a World Environment Organization (WEO).[79] Along with
nongovernmental actors such an organization could, these institutionalists believe, help embed environmental norms in the global political
economy (see chapter 8).

Most social greens and some bioenvironmentalists are far more critical of the WTO, often calling for a complete overhaul of the institution, or sometimes even its dissolution. Many social greens see the
WTO/GATT system as an instrument used by capitalist states and corporations to liberalize trade.[80] Moreover, they argue that it is secretive,
undemocratic, discriminatory, and antienvironmental. For this reason
they see genuine reforms using its existing framework as nearly impossible.[81] They praise the few international environmental agreements that
include possible trade sanctions.[82] But they also argue that these treaties
could go much further to restrict environmentally harmful trade. In addition, many social greens and bioenvironmentalists justify unilateral trade
actions as a sovereign right of states to protect their own (and the global)
environment.[83] Some also propose building alternative global institutions. Colin Hines of the International Forum on Globalization, for
example, advocates a General Agreement on Sustainable Trade (GAST).[84]
A GAST would not discourage trade, but rather it would encourage
states to treat domestic industry favorably, as well as allow states to give
preferential treatment to goods from other states that respect human
rights, labor rights, and the environment. It would also allow citizen
groups to sue companies that violate the agreement.[85] Other proposals
include an expansion of "fair trade"—that is, small-scale trade with
communities that pay fairer prices for goods and that can ensure

environmentally sound production. The fair-trade movement, though growing for items such as coffee, handicrafts, and bananas, is still a tiny proportion of global trade.[86]

Regional Trade Agreements—Opportunity for Greener Models?

If there is some difficulty reconciling the tensions between trade and the environment within the WTO, what are the prospects for regional trade agreements? Here we look at NAFTA, the European Union, and the Asia Pacific Economic Cooperation (APEC) as examples of regional trade agreements and how each grapples with issues of the environment.

NAFTA, negotiated in the early 1990s, is arguably a more environmentally friendly trade treaty than the GATT/WTO. It explicitly deals with environmental issues to a much greater extent than the GATT, specifically mentioning the environment in the preamble and main text of the treaty (in five of twenty chapters). Chapter 1 of the NAFTA text stipulates that the trade provisions embodied in the environmental treaties named in the trade agreement take precedence (it mentions the multilateral treaties CITES, Basel, and the Montreal Protocol, as well as four bilateral treaties). This exemption for these agreements is qualified, however. NAFTA parties must choose the least trade-inconsistent measure to meet their obligations under these treaties, regardless of whether that measure is more politically or economically feasible.[87] Chapter 11 of the NAFTA text also explicitly discourages pollution havens. That is, it states that it is inappropriate for parties to encourage investment by relaxing domestic standards for health, safety, or the environment. (The next chapter of this book examines the concept of pollution havens and the investment provisions of NAFTA more closely.) Other parts of the NAFTA agreement also mention the environment in more flexible and explicit ways than does the GATT, incorporating a more precautionary approach.[88]

There is in addition an environmental side agreement parallel to the measures in the NAFTA text: the North American Agreement on Environmental Cooperation (NAAEC). The purpose of the side deal is to foster environmental cooperation among NAFTA parties, to ensure that states comply with and enforce their own national environmental laws, and to establish a mechanism for settling environmental disputes. Parties are required under the side agreement to report periodically on the state of their environment. A new institution, the Commission on Environmental Cooperation (CEC), was created to oversee the NAAEC. It conducts studies on environmental performance and policies in the NAFTA countries; it also hears environmental complaints and settles disputes. If a party feels that one of the other parties to the agreement consistently fails to enforce its own environmental laws, it can bring a complaint forward to the CEC. The CEC is much more open than the WTO to citizen and NGO participation, and these groups can individually launch such complaints.[89] The side agreement also established the Border Environmental Cooperation Committee (BECC) and a North American Development Bank.

There were serious disagreements over the environmental dimensions of the NAFTA treaty during negotiations, including among environmental groups, particularly those in the United States. Groups drawing on social green and bioenvironmentalist analysis, including Friends of the Earth, Sierra Club, Public Citizen, and Greenpeace, were highly critical of most aspects of the agreement. In particular these groups opposed the agreement for what they foresaw as its negative environmental and development impact. They were critical of the negotiation process, too, which they felt corporate interests had effectively co-opted. A more moderate position, taken by groups following more of the reasoning of institutionalists and market liberals, including the National Resources Defense Council, the WWF, and Environmental Defense, more or less endorsed the way environmental concerns were brought into the treaty. They basically agreed that trade liberalization was a potentially positive force, or was at least going to occur regardless, and wanted to ensure that environmental safeguards were built into the treaty.[90]

It is still too early to fully assess whether NAFTA has been "greener" in practice than other trade agreements. On the positive side, in the 1990s there was no major decline in standards or enforcement in the United States.[91] Meanwhile Mexico's environmental laws have improved since NAFTA came into force, with levels of enforcement stable.[92] Serious environmental violations in the industrial region of Mexico near the U.S. border have also decreased by 72 percent, with over four hundred firms signing environmental-compliance action plans.[93] On the negative side,

Canada's post-NAFTA record has been less impressive.[94] So far, the CEC has also not accomplished much. The studies it has conducted on environmental quality after implementation of NAFTA have been criticized for methodology and issues chosen (e.g., agriculture rather than toxic waste). Critics further argue that NAFTA has devastated Mexican smallholder agriculture, altering land-use practices and polluting the countryside with toxic chemicals. Mexican and Canadian imports of toxic waste from the United States have increased since NAFTA. Mexico alone has seen a doubling of hazardous-waste imports since 1994. A report published by the Texas Center for Policy Studies indicates that exports of hazardous waste from the United States to Mexico increased from 158,543 tons in 1995 to 254,537 tons in 1999. Similarly, Canada saw its hazardous-waste imports rise from 383,134 to 660,000 tons in the same period.[95]

The case of the European Union's handling of the trade and environment tensions is somewhat different from that of NAFTA. The origins of the European Union go back to 1957 when the Treaty of Rome was signed, an agreement that sought to promote economic and political integration among European states (rather than simply trade liberalization).[96] The Single European Act (1986), the Treaty on the European Union (1992), and the Treaty of Amsterdam (1997) have reshaped the European Union in a way that promotes further political and economic integration, including the harmonization of environmental laws.[97] Over this period the European Union has developed and adopted EU-wide policies for environmental protection, and these are not seen to be secondary to economic goals.[98] A number of policy directives have been adopted regarding the environment, in an attempt to harmonize regulation across the European Union. These include policies regarding waste, air pollution, water, nature protection, and climate change, for example.[99]

The European Union's goal of political as well as economic integration has meant that the implementation of harmonized environmental laws has taken place parallel to, but separately from, economic integration. For this reason, there has not been much controversy over the harmonization of environmental laws across Europe and its economic implications. In fact, harmonization of standards has been considered important in facilitating trade in the region. Certain EU states, however—primarily those with weaker environmental standards to begin with, such as Greece, Portugal, Spain, and Italy—were more hesitant than others to adopt stringent EU-wide environmental policies. For this reason the European Union's environmental directives have some flexibility built into them, to allow for some degree of difference in requirements according to member states' economic and environmental situations.[100] Despite these differentials, however, the European Union is widely seen to be a successful case of upward harmonization of environmental laws within the context of enhanced economic integration.

The handling of trade and environmental issues in APEC has had a much different outcome than under NAFTA and in the European Union. It has been argued that although "sustainable development" has been a key policy goal of APEC, little to date has been done with respect to environmental cooperation in the region.[101] APEC is a broad regional grouping consisting of a number of countries that trade with one another across the Pacific Ocean, including those in East and Southeast Asia as well as North and South America. As such, it includes both advanced industrial countries (e.g., Japan, the United States, Canada, Australia) and a number of newly industrializing countries (e.g., the Philippines, Indonesia, Mexico, Chile) and countries in transition (e.g., China, Russia).[102] APEC was established in the early 1990s as a means by which to foster economic relations across the Pacific after the end of the Cold War. The aims of APEC are to facilitate trade, including trade liberalization, as well as to promote economic and technical cooperation, with the latter including environmental cooperation. Though these two goals are central to APEC, it has pursued them on separate diplomatic tracks. Moreover, the members of APEC have a particularly wide diversity in terms of environmental and economic conditions, making agreement in this area particularly difficult. As a result, there has been very little integration of trade and environmental issues, and little by way of strong environmental policy across the region has emerged.[103]

Conclusion

There is a growing consensus within the global policy community that trade and environment are mutually compatible—indeed, that both are essential for the health of the other. The institutionalist position on trade has grown stronger over the last decade, and today, the global policy goal adopted by most governments is to "manage" trade to foster *both* economic growth and better environmental conditions. This requires stronger national environmental policies and international environmental laws and norms. It also requires states to integrate environmental concerns into free-trade agreements. Yet the collective constraints of these rules must allow trade to still remain "free enough" so it can continue to power economic growth in all corners of the globe. It is critical, in this view, to avoid using trade as a mechanism for environmental protection, because this only distorts the efficient allocation and use of global environmental resources.

This does not mean the "fight" over trade is over. Market liberals still see the potential to go too far—and some feel this has already happened. That is, global environmental regimes will become yet another barrier to the accumulation of global wealth—and that, in turn, will impoverish all. Many bioenvironmentalists and social greens are far more skeptical than market liberals about the compromise of managed trade. They argue that it does not address the overconsumption or gross inequalities intrinsic to the global trading regime. It does little to restrain the destructive activities of multinational firms or to internalize the full social and ecological value into the prices of traded goods. It also does little to raise global environmental standards, and in many cases allows (and even encourages) states to keep domestic standards low (and can even create incentives for states to lower standards further). In short, social greens and bioenvironmentalists see global trade and free-trade agreements as little more than engines of unsustainable macroeconomic growth, creating ever-larger stress on the global ecological system.

Even many institutionalists feel there is a long way to go—although these concerns are more about particulars and less about the theoretical foundations of trade and the environment. Many institutionalists, for example, argue that under the WTO and NAFTA in particular, it is still far too difficult to ban trade in harmful products, especially on the basis of production or processing methods. Many note further that global trade rules do not set any minimum standards for environmental regulation—only ceilings, not floors. It is much easier under international trade law to contest another country's high environmental standards as unfair; it is far harder to contest weak environmental standards as unfair. Thus, the safer position for states is to weaken standards, reinforcing the race to the bottom. Finally, many institutionalists point out that the overlap of trade agreements with environmental clauses and environmental agreements with trade provisions creates unclear, and at times ambiguous, global trade and environment rules.

The norms of free trade and environmental trade clauses steer much of the environmental conduct of transnational corporations. In the next chapter we explore the environmental consequences of TNCs in more detail.

6

Global Investment and the Environment

The public often associates corporations with the world's highest-profile transnational firms: ones like General Electric, Citigroup, ExxonMobil, Wal-Mart Stores, Microsoft, General Motors, Procter & Gamble, Johnson & Johnson, and ChevronTexaco.[1] There are, however, tens of thousands of parent TNCs and hundreds of thousands of affiliates. Sales and profits drive these firms. Economic globalization, which is extending free trade, privatization, deregulation, and the amount and speed of global financial flows, is also enhancing the power of these firms. There are ever more opportunities for higher sales and higher profits in a borderless world market. A quick glance at the opportunities in China—with over a billion people and with rapid increases in domestic consumption—is illustrative. Now that China is a member of the World Trade Organization (WTO), TNCs are lining up to invest there, because there is potential for immense profits from such investments.[2]

Market liberals applaud these developments. Successful TNCs mean a healthy global economy. A healthy global economy means strong growth in national economies—in both rich and poor countries—which in turn translates into more state and corporate funds for better environmental management. TNCs, moreover, have the financial resources for sizable investments in developing countries, bringing new technologies (which are cleaner and more efficient than their predecessors) as well as higher management standards. TNCs are, in other words, the engines of sustainable development for both developed and developing countries. Institutionalists agree here, although they add that the profit imperative of firms means that in some cases the international community (through laws and financial incentives) needs to guide the actions of firms to ensure sustainable development.

As with trade, bioenvironmentalists and social greens hold very different views from market liberals and institutionalists on the environmental impacts of TNCs. Bioenvironmentalists see TNCs as engines of overproduction and overconsumption. TNCs extract, process, and export to rich countries the bulk of the globe's natural resources, creating deforestation, overfishing, overmining, and desertification. They manufacture, brand, and sell the bulk of the world's products, constructing a culture of consumerism with clever advertising. They pollute the earth's air and water with dangerous chemicals like dioxins, furans, polychlorinated biphenyls (PCBs), and DDT. And they transport and improperly dispose of garbage and hazardous waste, sometimes shifting it from rich to poor countries. Social greens largely accept the bioenvironmentalist evaluation of TNCs with respect to the environment. They add, however, that in the quest to maximize profits TNCs are responsible, too, for the inequality and exploitation of much of humanity. Billions of people now scramble for small sums of currency to survive, tilling increasingly unproductive land for export crops (e.g., coffee and nuts) and slaving away in perilous factories (e.g., to sew cheap clothes). The nutritional balance of subsistence farming is lost. So is the community well-being and cooperative environmental management of the past. TNCs are, in other words, the engines of injustice and exploitation of both nature and humanity.

This chapter begins with an overview of recent trends in the globalization of TNCs and foreign direct investment. It then examines the debate over environmental standards, addressing in particular the question of whether firms relocate to take advantage of lax environmental rules in developing countries. Following this is an analysis of the site practices of TNCs in the developing world. The chapter also examines the reaction of TNCs to pressures for environmental reforms, dividing the discussion into those who argue the corporate sector is "greening" and those who argue it is all "greenwash." It ends with a discussion of the role of TNCs in global environmental and investment governance.

Globalization and Transnational Corporations

Foreign direct investment (FDI) by TNCs is one of the main consequences and drivers of economic globalization. FDI entails ownership or investment in overseas enterprises in which the investor plays a direct managerial role.[3] A TNC by definition invests in overseas operations. Generally, a parent firm of the TNC holds minority or majority shares in numerous branches, subsidiaries, or affiliates in more than one country, creating complex managerial structures.[4]

The number of TNCs has increased markedly since the 1970s, growing from 7,000 TNC parent firms in 1970 to over 65,000 in 2002, with some 850,000 foreign affiliates around the world (see figures 6.1 and 6.2). Foreign affiliates now account for one-tenth of global gross domestic product (GDP) and intrafirm trade accounts for one-third of global exports.[5]

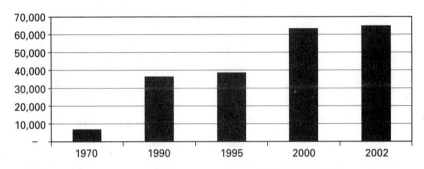

Figure 6.1
World total number of TNCs (parent companies). *Data source*: UNCTAD, *World Investment Report 1990, 1995, 2000, 2002.*

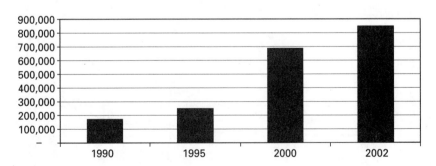

Figure 6.2
World total number of TNC foreign affiliates. *Data source*: UNCTAD, *World Investment Report 1990, 1995, 2000, 2002.*

World FDI net inflows have risen significantly since the 1970s, although much of the acceleration has occurred in the 1990s. In 1970 FDI net inflows were US$9.2 billion, growing to US$689 billion by 1998.[6] In 2000, the level of FDI inflows reached a record US$1.49 trillion before falling to US$735 billion in 2001 (see figure 6.3).[7] Not surprisingly, the total value of global inward FDI stock has been growing steadily over the last four decades: from US$68 billion in 1960 to US$3.2 trillion in 1996, and to US$6.8 trillion in 2001.[8] While this growth in FDI has been phenomenal over the past decade in particular, like trade it has not been evenly distributed. In 2001 developed countries accounted for 68.4 percent of FDI, with Western Europe, the United States, and Japan attracting 45.7, 16.9, and 0.8 percent respectively. Developing countries accounted for only 27.9 percent of FDI inflows in 2001, with 62 percent of those flows going to just five developing countries. Of the developing-country regions, the Asia-Pacific area and Latin America attracted 13.9 and 11.6 percent of global FDI inflows respectively in 2001, while Africa attracted a mere 2.3 percent. Central and Eastern Europe accounted for 3.7 percent of global FDI inflows in 2001.[9] While there is a link between global levels of FDI and global economic growth rates, that link is carried over to developed but not developing countries.[10]

The liberalization of investment rules over the course of the 1980s and 1990s has had much to do with the rapid pace of growth in FDI and TNCs in that period. Between 1991 and 2001, of some 1,393 regulatory

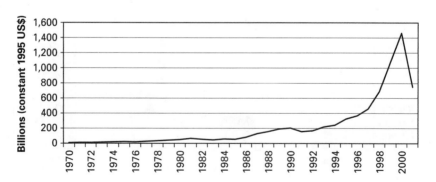

Figure 6.3
World foreign direct investments, net inflows (balance of payments). *Data source:* World Bank World Development Indicators: www.worldbank.org/data.

changes to domestic FDI rules, 1,315 (94.4 percent) were changes that created a more favorable environment for investment.[11]

Many global corporations have expanded their power through mergers or acquisitions over the past decade. Measured in 2001 U.S. dollars, the value of cross-border merger and acquisition activities by TNCs was US$151 billion in 1990 and US$1.14 trillion in 2000, falling to US$594 billion in 2001.[12] Between 1986 and 1990 cross-border mergers and acquisitions between TNCs grew at an average annual rate of 26.4 percent—in 2000 alone the rate was 49.3 percent.[13] The result has been increasingly large corporations. Indeed, in 1999 the value of the combined sales of the top 200 corporations was larger than the combined GDPs of all countries minus the 10 largest.[14] Globalization has also concentrated corporate power by facilitating control over specific markets. In the mid-1990s, the five largest electronics and electrical equipment companies shared 40 percent of the global market; in consumer durables, the five largest companies controlled 70 percent of the global market.[15]

Differential Standards: Pollution Havens, Industrial Flight, Double Standards?

Critics of TNCs often accuse them of intentionally locating their operations in areas with weak or poorly enforced environmental regulations. A core debate here is over whether highly polluting industries migrate to countries with lax environmental regulations. Is there an "industrial flight" of firms from countries that impose stringent environmental regulations? Do firms seek "pollution havens" in countries with lower environmental standards?[16] To follow these debates, it is essential to distinguish carefully between industrial flight and pollution havens. Industrial flight occurs when the *raising of environmental standards* in one country triggers the relocation of that industry to another country with lower standards. Pollution havens exist when two criteria are met: first, a country (usually, but not necessarily, a developing country) *sets its environmental standards at an inefficiently low level* (in economic terms) in order to attract foreign investment; and second, industry relocates to this country in order to save on pollution-abatement costs. A pollution haven is not simply a state with low environmental standards. Nor is it simply an area with lots of pollution. These situations, all agree, are common throughout the developing world. World Bank economist David Wheeler explains:

To qualify, the region has only to use weak environmental regulation as an inducement to investment by cost-conscious polluters. These polluting firms may be owned by either domestic or foreign investors; they may use new or re-located facilities; and they may produce for either domestic or foreign markets. What really counts for the pollution havens debate is neither ownership nor market location, but the willingness of the host government to "play the environment card" to promote growth.[17]

Pollution havens and industrial flight are thus two sides of the same coin: importantly, both concepts seek to capture the implications of differential environmental standards (and hence, costs) between different countries.

The debate over whether pollution havens and industrial flight are real phenomena began in the 1970s. It arose out of the fear that higher environmental regulations in the United States might drive industries, and with them jobs, to developing countries with lower environmental regulations. This concern remains strong even today. But it is paralleled by a concern about the environmental and health implications for countries that host dirty industries.[18] There is particular concern with the state of the environment in the developing world, because conditions there continue to deteriorate as those in the industrialized world improve.[19] Again, though, it is critical to note that this does not confirm the presence of industrial flight or pollution havens. We now turn to the evidence and arguments for and against the existence of these phenomena.

Differential Standards Are Inevitable

There were a number of studies in the 1970s and 1980s to test the underlying hypothesis connecting pollution havens and industrial flight—that is, the idea that firms relocate in response to changes in environmental regulations. Coming mainly from a market-liberal perspective, these studies were primarily econometric and drew on pollution emissions and emission-control cost data for U.S. manufacturing firms. Most of the studies found that the probability of firms relocating to developing countries in order to avoid more stringent environmental regulations in rich countries was low.[20] One of the main reasons cited for the "elusiveness" of pollution havens and industrial flight is often simple: costs like labor and technology are far higher than environmental costs and therefore a shift in environmental regulations is not enough to encourage firms to relocate on those grounds alone. Pollution-control costs rarely constitute more than 2 percent of sales.[21] There are other factors as well. Some industries—such as electricity generation—need to remain close to their market, and it may be impractical to relocate. Also, for some industries stable and predictable standards matter more than compliance costs.[22]

The intensity of pollution is undeniably rising in some poor countries. Yet market liberals like David Wheeler do not believe this is connected to TNC reactions to regulatory differences between rich and poor countries. Wheeler in fact found that over the last decade air quality had improved in the three largest recipients of FDI in the developing world—Brazil, China, and Mexico.[23] The decline in investment in "dirty" industries in industrialized countries between 1960 and 1995 has no doubt corresponded with increasingly tight environmental regulation. And there has no doubt been a rise in many of the same pollution-intensive industries in the industrializing world. Yet market liberals maintain that the cause is shifts in *domestic* production and consumption patterns— often the products from "dirty methods" are for home use, not for export—and not shifts in TNC investment to take advantage of regulatory differences. There is, in other words, no "race to the bottom" as a result of industrial flight from rich countries or the pull of pollution havens. Moreover, in the rare instance when pollution havens arise, these are relatively short-lived phenomena. Economic growth soon provides the means for a more adequate environmental response such as improved regulation and a rise in investment in cleaner production processes.[24]

For market liberals, then, differential standards are not causing an influx of foreign investment into developing countries. Some have even argued that lax standards deter FDI.[25] Environmental problems arise mainly as part of domestic industrial growth (as we saw the EKC predict in chapter 4), not differential standards. For most market liberals, differences in environmental standards between countries are normal, or at least inevitable. In their view, as we saw in chapter 5, it is unrealistic to expect all countries to maintain identical standards and ecological conditions. Different countries quite naturally have different pollution-absorption capacities, both politically and physically.[26] Such differences can even give countries useful absolute or comparative advantages for production and trade. A 1991 World Bank memo, under the signature of Larry H. Summers, former vice president and chief economist of the World Bank, former U.S. Treasury Secretary in the Clinton administration, and current president of Harvard University, bluntly explained the logic:

The measurement of the costs of health-impairing pollution depends on the foregone earnings from increased morbidity and mortality. From this point of view a given amount of health-impairing pollution should be done in the country with the lowest cost, which will be the country with the lowest wages. I think the economic logic behind dumping a load of toxic waste in the lowest-wage country is impeccable and we should face up to that.[27]

Market liberals contend that differential standards occur alongside the transfer of clean technologies and better management to developing countries. In developing countries, they argue, even TNCs in dirty industries are more likely than domestic firms to use more sustainable technologies and more elaborate environmental policies.[28] Why do TNCs perform better than local firms (which are often much smaller)? Most importantly, perhaps, TNCs have the financial and organizational resources to access more sophisticated technologies and managers. There are other reinforcing factors as well. TNCs sometimes keep overseas standards relatively high as part of a risk-management strategy, so sudden environmental reforms or lawsuits do not disrupt corporate activities. They sometimes raise environmental standards overseas to "voluntarily" satisfy NGOs, international organizations, and consumers. TNCs may also employ higher standards and cleaner technologies to gain a competitive advantage in production (new technologies are generally more efficient and produce higher-quality products), and this efficiency is reinforced when they use the same practices in all of their manufacturing plants, regardless of location or differential standards.[29]

Market liberals tend to argue that smaller and local firms in the developing world are in fact more polluting than larger and multinational-affiliated firms. This is in large part because they lack funds to install cleaner production technologies. Domestically owned firms also tend to be more secretive and less accountable to both public and "outside" pressures. In addition, smaller firms are more dispersed, and for this reason it may be harder to ensure their compliance with environmental regulations. Locally owned firms may also be more polluting because incentives to be "cleaner" are lower where land ownership and rights are not clearly delineated, as is the case in much of Southeast Asia.[30] Market liberals see state-owned local firms in the developing world as particularly prone to be heavy polluters.[31]

While they point out these differences between the environmental performance of TNCs and domestic firms, market liberals *do not* argue that the overseas standards of TNCs are the same as those in their home country. Rather, they stress that the environmental performance of TNCs is often better than that of both domestic firms and the minimum standards of host-country laws. To further improve environmental performance of both domestic firms and TNCs, market liberals call for market-friendly measures like public disclosure of pollution emissions and pollution fines. They highlight cases in the developing world, such as in Indonesia and the Philippines, where public disclosure of emissions has had an impact on the environmental practices of firms.[32]

Differential Standards Cause Environmental Harm

Others are far more skeptical of the impact of foreign direct investment on the natural environment. Social greens are perhaps the most critical, although bioenvironmentalists and institutionalists share some of their concerns. Most see pollution havens and industrial flight as real or potential threats. Indeed, the most environmentally damaging industries tend to have a high concentration of TNCs as the main investors and operators.[33] Some of the worst offenders are global extractive TNCs that mine natural resources—loggers, oil drillers, and miners. This pattern of foreign direct investment in high-environmental-impact industries has afflicted both developed and developing countries, although the latter are now receiving more and more of this investment. The evidence presented by social greens and bioenvironmentalists, though, relies less on the statistical data of market liberals and more on in-depth case studies. Much of it, market liberals retort, is mere anecdotes. But for social greens in particular this evidence allows for analysis of the impact of TNCs on real people—on their individual lives and the well-being of communities, notions not easily captured by econometric surveys of data.

Social greens and bioenvironmentalists are concerned about the rise in the share of highly polluting industry in developing countries, and accuse TNCs of following what they call "double standards"—that is, when a firm applies one set of standards in their operations in one country, and a different set of standards in another.[34] Usually this means more lax standards are applied in TNC operations in developing countries, while higher standards for the same production processes are adhered to in the home country. Several UN studies have confirmed that TNCs operating in developing countries conform to standards lower than home-country standards.[35] As mentioned earlier, market liberals do not deny that this occurs—they just argue that this is still a better situation than following host-country standards. Social greens in particular see this practice as unjust, exploitative, and dangerous.

Social greens highlight the case of the American TNC Union Carbide in Bhopal, India, the site of the worst industrial accident in history, as a clear case of double standards and environmental injustice.[36] The Carbide Bhopal plant produced pesticides. Union Carbide owned 51 percent; the Indian government owned the rest. Many claim that Union Carbide was aware in the early 1980s that this factory had much more lax environmental, health, and safety standards than a similar pesticide plant owned by the same company in West Virginia (which produced the identical pesticide).[37] Little was done to upgrade the Indian plant. Then, in the middle of the night on December 4, 1984, there was a leak of poisonous gas at the Bhopal plant that spread over the town. Eight thousand people were instantly killed, and several hundred thousand more were injured.[38] Union Carbide blamed the tragedy on the Indian government for not conducting better checks on the plant, and argued that sabotage likely caused the accident. But critics blamed it on the negligence of Union Carbide, arguing that it was not an accident but rather a "disaster waiting to happen."[39] The company was taken to court and was forced to pay compensation to many of the victims, though many others received nothing for their suffering and those that were given funds only received a small settlement.[40] Union Carbide pulled out of India 10 years later, and the company was recently bought out by Dow Chemical. An investigation by the *New Scientist Magazine* claims the U.S. headquarters was responsible for the plant's design, and that Union Carbide had scaled back its investment in safety technologies at Bhopal in the early 1980s.[41]

The maquiladora manufacturing plants in Mexico are another case of TNC environmental abuse in a developing country. About 2,000 American companies, including General Electric, Ford, General Motors, and Westinghouse, have moved their assembly plants over the border to Northern Mexico, avoiding environmental laws in California regulating toxic emissions. There are in fact reasonable environmental laws on the books in Mexico. But the Mexican government does not enforce them effectively. In the 1960s and 1970s most of the maquiladora factories were in sectors such as garment assembly and were not highly polluting. Over the course of the 1980s, however, there was a dramatic change in the composition of these factories. The main sectors are now chemicals, electronics, and furniture, all of which are highly polluting.[42] By the early 1990s nearly 90 percent of maquiladoras used toxic materials in their production. One reason for these changes was the increase in the quantity of investment in these sectors since the 1970s.[43] For example, from 1982 to 1990 investment rose twentyfold in maquiladora chemical factories after a tightening of environmental regulations on the industry in the United States.[44] Environmental studies of conditions near these plants over the past two decades have revealed polluted local rivers, soils, and water supplies. Workers have acute health problems, and there has been an increase in birth defects. At fault, critics claim, are corporations that are ignoring environmental regulations.[45]

For social greens and bioenvironmentalists the ecological performance of TNCs in these cases is typical of how TNCs generally exploit the peoples and environments of developing countries.[46] Social greens stress the injustice and environmental harms TNCs cause, while bioenvironmentalists tend to focus mainly on their ecological impacts. Both argue that economic globalization means the smallest cost differences are increasingly influencing corporate decisions on where to locate, and on how these corporations will operate once they have relocated. It also means that governments increasingly hesitate to regulate strictly for fear this will drive firms away. Globalization, then, is not diffusing cleaner technology globally as the market liberals argue, but allowing more and more TNCs to exploit remote habitats and peoples.

Bioenvironmentalists see all corporations—usually in the context of capitalism, profit maximization, and overconsumption—as potential sources of ecological problems. Social greens see TNCs as unconnected to local communities, which makes them more likely to exploit workers and the environment. For them, local firms are innately better environmental managers. Local firms are more oriented toward local needs. They tend to be labor rather than capital intensive, and generally function on a smaller scale than TNCs. They also tend to produce goods more appropriate to the needs of the people in the nearby region. Of course there are many exceptions, as in China, where small-scale manufacturing enterprises account for a significant proportion of pollution. Nevertheless, in many developing countries, TNCs produce the bulk of goods and thus pollution.

Expanding the Debate

There is also a growing position among critics of the market-liberal view of TNCs that the debate over differential environmental standards has been too narrow and too polarized, making it difficult to move forward with concrete policy steps. The economic literature on pollution havens has dominated the debate. It defines "dirty industry" in narrow terms and in doing so fails to adequately encompass location decisions of all polluting firms that operate globally. It tends to rely heavily on data on pollution emissions or on expenditures for emission controls by manufacturers to determine which sectors are highly polluting and dirty. This focuses too much on manufacturers, critics argue, and ignores some of the world's worst polluters and degraders, such as the hazardous-waste and natural resource firms.[47] There may be havens of environmental degradation, rather than simply industrial pollution, but such a narrow framework ignores them.[48] Another problem with the econometric studies is their focus on expenditure on emission controls as the only environmental cost faced by firms. This tends to underestimate total environmental costs.[49] Other environmental costs, though perhaps harder to measure, are equally important. These include costs related to public relations, liability, insurance, and certification of voluntary standards such as Responsible Care or ISO 14000 (discussed later).

Institutionalists in particular are somewhat critical of the market-liberal search for statistical proof of pollution havens (defined narrowly). The global community, they argue, needs to take a broader approach, one that focuses on policies to remedy the inequities and detrimental impacts of differential standards. Such an approach needs to explore incentives—real or perceived—that affect state policies and corporate behavior. These incentives must be as much political as economic. They need to address issues like the possibility of a "regulatory chill"—a problem related to pollution havens but one most market-liberal analyses of pollution havens ignore. The phrase "regulatory chill" describes instances where countries fail to raise environmental standards above the status quo for fear that investment will either leave or not enter the country. Whether firms actually respond to this failure to raise standards is not the core issue. Rather, the point is that the fear that firms will act can affect the stringency of regulations in both rich and poor countries.[50] A regulatory chill is almost impossible to quantify. There are studies, however, that indicate that the fear of possible corporate reactions does affect government decisions on whether to strengthen environmental regulations. In both the United States and the European Union, for example, industry groups successfully convinced governments to not impose stricter climate-change regulations by stressing competitiveness concerns.[51]

Those who call for a broader debate tend to advocate policies that are "win-win" even in the absence of hard evidence that pollution havens exist. All groups can agree on this broad point, but it is far harder to reach a consensus on the nature and extent of restrictions on corporate behavior. Those leaning toward market-liberal and institutionalist views primarily advocate aid from rich to poor countries, support for public access to corporate information, stronger regulations, and more cost-effective ways to reduce pollution. Some also argue that the International Monetary Fund and World Bank should account for risks of more intensive pollution from structural adjustment programs.[52] Others stress the need for even stricter actions, such as a global investment agreement that would put rigorous environmental controls on foreign direct investment by TNCs (discussed later).

TNCs and Site Practices

A second major strand of literature on corporate investment and the environment centers not so much on whether firms relocate to take advantage of lower environmental costs, but instead evaluates the actual environmental practices of transnational corporations. Often, authors concentrate on a simple question: Do corporate operations harm the environment? Critics of corporations have dominated much of the literature here, often describing illegal and destructive activities of specific industries in specific regions.[53] Research institutes like the World Resources Institute and NGOs like Project Underground, the Rainforest Action Network, and Global Witness in Cambodia document these activities, with their findings then distributed worldwide through the Internet and free publications.[54] The literature on natural resource firms and manufacturers tends to generate somewhat different criticisms and debates. So does the literature on the site practices of firms in developed countries (which often have better environmental practices than in the past) versus ones in poor countries (which often have lower operating standards and less transparency).

To illustrate this body of literature, we briefly examine three natural resource sectors (logging, mining, and oil) and two manufacturing sectors (chemicals and electronics) in the developing world. This literature tends to reflect the views of bioenvironmentalists and social greens, because market liberals and institutionalists tend to presume that correct markets or correct institutions will solve these problems. Social greens and bioenvironmentalists see the problem as more entrenched and as arising from the nature of capitalism and corporations themselves. For them, capitalism, the enshrining doctrine of profits and greed, creates marauders and plunderers.[55] It is little more than a flashing neon sign: "Earth for Sale."[56] Globalization is enabling corporations to "Rule the World."[57] It is creating a "Corporate Planet."[58] Firms are becoming larger, more powerful, and less accountable, combing the world for cheap labor and "limitless" environmental resources.

Irresponsible and illegal corporate loggers are seen by social greens and bioenvironmentalists as one of the main causes of forest degradation in the tropics (which opens the forest to fires and farmers and triggers the process of deforestation).[59] While bioenvironmentalists also stress population growth in conjunction with corporate behavior as key causes, social greens focus almost exclusively on the role of corporations in the global capitalist framework. Loggers routinely skirt environmental rules, often relying on state officials for access to sites and protection from prosecution. It is common—indeed it is often necessary in order to stay in business—to bribe enforcement officers and customs officials. Numerous studies have found that few firms, if any, actually practice sustainable logging (that is, in a way the forest can regenerate so loggers can harvest the same amount again sometime in the future—say in 50–80 years).[60] A glance at the Asia-Pacific region confirms the terrible consequences of logging. TNCs from Japan, Malaysia, and Indonesia have been at the center of the tropical logging industry over the last four decades. Japanese general trading firms like Mitsubishi Corporation have supplied funds and "guaranteed" markets.[61] Malaysian and Indonesian firms have often conducted or supervised actual logging operations. Natural forests still blanketed most of the Asia-Pacific area in the middle of the twentieth century. Now, forests cover just three-quarters of Melanesia, about half of Indonesia, Laos, and Cambodia, and just under half of Malaysia and Burma. The situation is even worse in Thailand (one-quarter) and the Philippines (one-fifth). The extent of forest degradation is an even greater problem. Around 95 percent of Asia's frontier forests—areas large and pristine enough to still retain full biodiversity— are already lost.[62] The Philippines is effectively logged out (it has been a net importer of tropical timber since the 1990s). The rest of the Asia-Pacific region will, at recent rates and under current logging practices, follow the Philippines in 10–20 years.[63]

Transnational mining companies have had a similarly poor record over the past decade, and have been targeted by environmental groups seeking to raise public awareness of the problem as well as to raise the environmental and social standards in the industry. Because most of the easily accessible minerals in the world have already been mined, mining TNCs are forced to search far and wide for new deposits, and are increasingly setting up operations in developing countries to extract minerals that are more and more remote and difficult to obtain. Because of this, risks in the industry are growing. And because of these risks, mining companies are increasingly seeking investment insurance and guarantees from multilateral development banks and export credit agencies (see chapter 7).[64] But such agencies place few environmental regulations on these operations. There have been numerous recorded accidents and scores of cases of environmental pollution as well as ecological and social disruption, including negative economic, health, and social impacts on women, resulting from mining operations over the past few decades.[65] Deforestation and toxic leaching of mine tailings and dangerous chemicals into waterways and the soil are just a few of the many environmental problems associated with mining. Mining accidents and environmental destruction at mine sites—such as the burst tailings ponds in Guyana in 1995 and in the Philippines in 1996—commonly occur because corporations do not take adequate precautions.[66] Often, miners displace indigenous peoples, and in some cases there are serious human rights abuses as TNCs make deals with host governments to suppress resistance. Activists have highlighted major ongoing environmental and human rights problems—for example, around the Grasberg mine in Indonesia and the Bougainville mine in Papua New Guinea.[67] While industry has taken some steps to improve its environmental performance, activists remain skeptical of their motives.

Oil exploration and extraction have similar negative environmental consequences. Oil is at the core of the global energy system, and thus of global development. According to Project Underground, an NGO focused exclusively on raising awareness about the oil and mining industries, the oil industry spends about US$156 billion per year seeking new oil and gas reserves.[68] As demand rises and reserves dwindle, the net has been cast over a larger and larger territory, including in the developing world. As with mining, TNCs are the main players in oil and gas exploration globally, and they often seek funds to support their ventures from export credit agencies. The oil TNCs with markets in Europe, Japan, and the United States can swamp economies in the developing world. In Ecuador, for example, the corporate push for construction of oil pipelines has led to a dependence on oil revenues. Because oil prices on world markets are highly unstable, the over-reliance on oil revenues drives a boom and bust cycle in the domestic economy, contributing to increasing levels of external debt as well as rises in poverty rates.[69] Activists have charged transnational oil companies operating in the developing world—from the Western Amazon to West and Central Africa to Southeast Asia—with human rights and environmental abuses. The environmental impacts of oil exploration are many, and include contamination of waterways from oily water wastes and toxic drilling muds that are stored in ponds similar to tailings ponds. Oil spills and gas flares also pose a constant threat to communities near oil drilling as well as during the cross-country and maritime transportation of oil.[70] As with mining, roads and settlements near exploration and drilling sites also trigger deforestation.[71] Human rights abuses of local peoples are also common.

Many were outraged in 1995 when the Nigerian government executed Ken Saro Wiwa and nine other activists for protesting the environmental impact of Shell's oil extraction in that country.[72] Social greens see these cases as clear examples of enclosure and the use of state violence to suppress indigenous groups (as discussed in chapter 4).

The site practices of hazardous manufacturing industries have also drawn the ire of critics. TNCs in the chemical industry in particular are frequently targeted. The chemical industry has increased its production in Asia and Latin America over the past decade.[73] Many of the chemical TNCs operating in developing countries produce highly toxic substances and wastes. Production of dangerous pesticides by TNCs and their affiliates, for example, is widespread in developing countries. In some cases, use of the chemical is highly restricted in industrialized countries but not in developing countries. Dursban, for instance, is a neurotoxin and suspected endocrine disrupter, yet Dow Chemical has continued to produce it in South Africa.[74] Transnational chemical companies are also engaged in the production chemicals and plastics that contain chlorine, a highly toxic substance. This includes production of polyvinyl chloride (PVC), a commonly used flexible plastic. When incinerated, chlorine releases dioxins and furans, some of the most hazardous chemicals known to humans.[75] Activist groups, seeking to discontinue unsafe practices, have documented the contamination of land and rivers with toxic wastes from illegal waste disposal and industrial accidents linked to chemical production in the developing world.[76] In some cases, governments have ordered firms to stop production in response to environmental and health complaints.

The activities of TNCs in the high-tech electronics sector, too, have been subject to scrutiny and criticism by activist groups. TNCs in this sector have also increased their investment in Asia and Latin America, but their environmental performance has not been as state of the art as the technology they are producing. The production of electronic goods and components—such as silicon chips, semiconductors, and computers—involves hazardous materials, such as arsenic, benzene, cadmium, lead, and mercury, all known carcinogens. These toxic materials are not only a threat to workers in this industry, but often end up in waste that pollutes land and waterways (local waste contractors hired by TNCs commonly dispose of this waste improperly). The manufacture of silicon chips and semiconductors also requires an enormous amount of water and energy, used for cleaning the products and keeping production sites clean.[77] In addition, the disposal of high-tech products at the end of their life is also problematic, especially because of their toxic components. While there are some recycling programs for old computers, cell phones, and other high-tech items, a significant proportion of those collected in North America for recycling end up being exported to developing countries, particularly China, India, and Pakistan. Serious environmental harm can occur when countries without strict regulations recycle such items. Not only is there exposure to the heavy metals in these items, but the burning of PVC-coated cables to reclaim the copper is highly polluting to the air and waterways.[78] Because the high-tech sector is one of the fastest growing in the global economy, critics argue that these problems are only likely to increase in the future unless action is taken now.

Greening or Greenwash?

In response to criticism of their poor environmental practices in the late 1980s and early 1990s, particularly in developing countries, many global firms began to "green" themselves.[79] With a lack of outside forces to regulate them, they have embraced self-regulation via voluntary initiatives. This enthusiasm by industry for self-regulation is driven in part by an attempt to improve their public image, and in part by economic factors. Market liberals have hailed the economic benefits of greening as proof that the market is self-correcting and that firms can deal with environmental problems. Social greens and bioenvironmentalists, however, are skeptical of the sincerity and efficacy of these efforts. Institutionalists on the other hand take the middle ground in this debate.

Greener

Throughout the 1980s and in particular in the run-up to the Rio Earth Summit in 1992, many firms began to take environmental concerns more seriously in their operations.[80] Responsible Care, an environmental and safety code delineating practices for the chemical industry, was established in 1985 following the accident at Bhopal. In 1990 the Global Environmental Management Initiative (GEMI), set up by the International

Chamber of Commerce (ICC), was charged with implementing the ICC business charter for sustainable development.[81] Industry also played a key role at the Rio Earth Summit in 1992 (see below). Indeed, Agenda 21 sanctioned the use of voluntary initiatives as a key strategy to promote sustainable development. Following Rio there was a flurry of such voluntary codes established by industry. The ISO 14000 environmental management standards of the International Organization for Standardization were established in the mid-1990s (see box 6.1), as well as a host of other "best-practices" codes in the resource-extraction sectors.[82] The UN's Global Compact with industry, launched in 2000, encourages corporations to follow nine principles of social and ecological responsibility.[83] At the 2002 Johannesburg Summit, industry again played a key role in promoting voluntary initiatives (see below). Through these efforts, industry has tried to keep the focus on voluntary corporate responsibility that includes environmental awareness—their preferred way of addressing environmental issues.

For market liberals, adherence to voluntary initiatives such as ISO 14000 promises only positive outcomes for the environment. Industry is seen as the best actor to set up such standards because it is most familiar with corporate possibilities and constraints. Market liberals prefer such an approach because it shifts the burden of regulation from the state to the firm, which can monitor environmental performance much more efficiently (important for "eco-efficiency"; see below). Moreover, adherence to global standards helps ensure that even when state enforcement is weak (common throughout the developing world), firms will still abide by local regulations. Global standards like ISO 14000 also encourage firms to adopt cleaner technologies, rather than focusing on cleaning up after the fact, because firms consider environmental impacts throughout the decision-making process. Such standards are intended to spread good environmental practices up and down the supply chain, as well as across borders, because certified firms are required to encourage suppliers to become certified.[84]

Industry coined the term *eco-efficiency* to refer to its new environmental awareness and the benefits this awareness would bring. By integrating environmental concerns into business practices, firms can expect economic as well as environmental benefits. These include lower costs for environmental cleanups, reduced capital costs due to lower risks,

Box 6.1
ISO 14000 environmental management standards

The ISO 14000 environmental management standards are perhaps the most widely recognized global-level voluntary initiative on the part of industry. These standards were developed in the early 1990s under the auspices of the International Organization for Standardization (ISO), directly following promises made by industry to establish voluntary initiatives at Rio. The ISO 14000 standards are management standards, meaning that they encourage firms to establish a management system that improves their awareness by setting their own goals for environmental improvement. The ISO 14001 standard, published in 1996, is the only certifiable standard in the series. To seek certification, a firm must provide an environmental statement showing it intends to comply with all applicable environmental laws in the jurisdiction in which it is located. It must commit to the prevention of pollution as well as to continually improve environmental management. It must adopt a management system that ensures that it conforms to its own environmental policy statement. Further, firms should encourage suppliers and contractors to establish their own environmental-management systems to conform to the ISO 14001 standard. By 2002, some 49,000 firms in 118 countries had gained certification to the ISO 14001 standard.[1] With a growing number of firms in both developed and developing countries adopting these standards, adherence to the standards will increasingly become a de facto condition for conducting business in the global marketplace.

Critics are concerned about whether the ISO 14000 standards will really make a difference in terms of environmental performance.[2] ISO 14001, for example, stresses that TNCs should comply with all environmental laws in the country in which they are operating. This differs substantially from Agenda 21, which calls on TNCs to follow home-country standards. In this way ISO 14001 allows for differences in standards between countries, but it may not do much to improve standards in developing countries. At the same time, some businesses are attempting to use ISO 14000 and other voluntary industry codes as a deliberate attempt to head off more stringent regulation. Industries in a number of countries are pressing their governments for some form of regulatory relief, such as more lenience for ISO 14001–certified firms during monitoring of environmental regulations. For example, the United States, Argentina, South Korea, and Mexico have adopted measures that take ISO 14001 certification into account in the monitoring and enforcement of regulations.[3]

1. ISO, 2003.
2. Krut and Gleckman 1998; Clapp 1998; Kimerling 2001.
3. Speer 1997, 227–228; Finger and Tamiotti 1999.

and increased market share secured by an enhanced image.[85] At the same time, economic efficiency is thought to have environmental benefits. Firms that are efficient generate less waste and incur fewer environmental costs. Efficiency and environmental protection then are seen to be synergistically beneficial. The push for eco-efficiency is a way to gain a competitive advantage in the marketplace, and firms that fail to adopt such principles could ultimately be pushed out of business. In this way environmental stewardship is viewed as making sound business sense.

Today, analysts like the ecological modernization theorists in the institutionalist camp argue that environmental concerns are being institutionalized—that these are becoming part of institutional structures and decision-making processes.[86] Drawing on experience from Europe, these thinkers see a restructuring of the capitalist political economy as taking place so that both firms and governments will view environmental management not as a cost, but as a business opportunity.[87] In this atmosphere, firms are increasingly going "beyond compliance," internalizing environmental norms and practices and thus seeking to intentionally do more than the law requires with respect to environmental practices.[88] Some prominent European oil companies, for example, have come out in favor of the Kyoto Protocol, beginning the process of moving their energy interests away from fossil fuels and toward alternative and renewable energy sources.[89] Some TNCs in the chemical industry are also seen to be "exporting environmentalism" by taking their new internalized environmental practices and codes of conduct abroad to their operations in developing countries.[90]

Greenwash

Undeniably firms everywhere now incorporate environmental language into public statements and official corporate policies. Critics of corporations, including most social greens and many bioenvironmentalists, label this as "greenwash," a phenomenon where a company tries to convince consumers and shareholders that it is environmentally responsible, where the purpose is more about image than substance.[91] An example would be a company that was instrumental in producing chemicals that destroyed the ozone layer now taking credit for "protecting" the ozone because it no longer produces these chemicals. Or a company that takes waste from one polluting process and uses it as a raw material in another, labeling it "recycling." Or a company that claims to be "green" but at the same time lobbies (or funds lobby groups) against environmental regulations.[92] Some observers further argue that companies are working hard to weaken environmental agreements or subvert environmental campaigns[93] or that firms are trying to "hijack" environmentalism to serve their own purposes.[94] Greenpeace took up the issue of greenwashing just before the Rio Earth Summit and published the *Greenpeace Book of Greenwash* to distribute at the event.[95] Groups like Greenpeace and CorpWatch target the large industry "leaders" they feel are advertising falsely and regularly issue "Greenwash Awards" to such companies. Activists Jed Greer and Kenny Bruno, in a book published by the NGO Third World Network, argue: "The reality is that TNCs are not saviors of the environment or of the world's poor, but remain the primary creators and peddlers of dirty, dangerous and unsuitable technologies."[96]

Social greens in particular argue that corporations are able to take the green high road mainly because they have managed to redefine sustainable development in a way that fits their own existing practices. Sociologist Leslie Sklair explains that "sustainable development was seen as a prize that everyone involved in these arguments wanted to win. The winner, of course, gets to redefine the concept."[97] As Sklair sees it, corporations won that argument in the 1990s and have redefined the concept of sustainable development in a way that suits their own interests. In doing so, companies have felt free to advertise themselves as being pro-environment, even when some of their practices are not. In many cases major corporations spend more on publicizing their green credentials than they do on the actual green programs they are advertising.[98] Critics argue that voluntary corporate responsibility (the industry term) should give way to corporate accountability, which would not just rely on promises to be responsible, but would also hold corporations accountable for their actions.[99]

Some of the more specific arguments made on this theme are that market liberals and institutionalists are conflating economic efficiency and environmental protection, which critics see as setting a dangerous precedent. In fact, nearly all of the examples of businesses adopting cleaner production strategies have resulted from government regulatory pressure, and did not emerge from within business itself as an efficiency-gaining measure.[100] Some have also pointed out that "green investment" touted by business is merely a "cleanup" rather than "clean." In short, critics claim industry is just changing labels rather than truly changing course.

TNCs and Global Governance for Investment and the Environment

At the same time that a fierce debate is raging over the environmental performance of TNCs, there is an emerging debate on the role of TNCs in the formation of global environmental governance, as well as on the implications for the environment of global rules that seek to regulate TNC activities. We discuss these issues in the following pages.

TNCs, through a number of different channels, have played an important role in the creation of global environmental governance. As mentioned in chapter 3, TNCs traditionally lobbied domestic governments in an attempt to influence the formation of global environmental rules, as a way to protect their own interests.[101] Increasingly, however, industry players are also lobbying policymakers directly at the international level, attending negotiations over specific environmental treaties as well as larger environmental conferences, as mentioned earlier. Compared to the 15-minute intervention made by industry representation at the Stockholm Conference in 1972, industry groups at both the 1992 Rio Earth Summit and the 2002 Johannesburg Summit put enormous effort into their involvement in these large summits.[102] Maurice Strong, the secretary-general of the Earth Summit, had encouraged his friend Stephan Schmidheiny of the Business Council on Sustainable Development (a corporate lobby group founded in 1990 that comprised some forty-eight TNCs) to take an active role.[103] The International Chamber of Commerce (ICC) was also active at the Rio Summit; it formed the World Industry Council on the Environment (WICE) in 1992 to provide industry follow-up on Rio. In 1995 the two groups merged to form the World Business Council for Sustainable Development (WBCSD). The Business Action on Sustainable Development (BASD) was formed in 2001 as a coalition of business groups (primarily a joining of forces of the ICC and the WBCSD, and comprising some 161 TNCs) for the chief purpose of lobbying at the Johannesburg Summit.[104] The main message put forward by industry at both of these gatherings, clearly within the market-liberal camp, was that voluntary self-regulation organized by industry, rather than outside regulation of TNCs, was the most effective and efficient way to promote sustainable development.[105]

The "structural power" of TNCs is another significant influence over global environmental governance. This refers to the ability of TNCs to influence the formation and functioning of governance through their dominant position in the global economy, which allows them to shape mainstream ideology and state-policy formation. It is extremely difficult to measure this type of corporate influence over states and global institutions, but many scholars stress its importance for understanding global environmental-policy outcomes. This type of analysis comes mainly from social greens who draw on a historical materialist perspective—in particular on the ideas of Gramsci. These scholars outline the ways a "bloc" of actors—consisting not just of transnational corporations, but also of certain states and key intellectuals who serve the interests of industry—influence the dominant ideology and discourse on sustainable development.[106] This influence is present in the role corporate actors play in defining sustainable development (as mentioned earlier) as well as in the negotiation of global environmental agreements.[107] The current era of increasing global economic competition, for these analysts, has meant that many states pursue domestic policy outcomes acceptable to corporations in order to keep or attract investment in their country.

Transnational corporate actors also influence global environmental governance through their participation in other international forums important to global environmental policy, such as through the setting of industry-based initiatives, as discussed previously. The ISO 14000 environmental management standards, for example—while initially designed as a voluntary set of standards—are now recognized as legitimate standards by the World Trade Organization. In this way they have become part of global structures of environmental governance. Some critics have expressed concern about this development, because the drafting process was dominated by TNCs with only minor input from governments and environmental groups.[108]

At the same time that industry players may influence the process of global environmental governance, intergovernmental investment rules also influence the actions of these firms, including ones not specifically designed to address environmental issues. Global and regional trade agreements with investment provisions, as well as agreements within the OECD, have had important implications for investment and the environment. Here we discuss the significance for TNCs of some of these investment rules.

One of the best known sets of investment rules with ramifications for the environment is chapter 11 of NAFTA, which sets out investors' rights under this trade agreement. The main defining features of this chapter are that it calls for governments to provide national treatment to foreign investors, and it allows corporations to sue governments for compensation when they believe their investments have been expropriated or when actions "tantamount to expropriation" harm their profits.[109] It sets up an investor-to-state dispute-settlement mechanism to handle such suits when they are brought forward. NAFTA is the first trade agreement to include such explicit rights for corporations. While chapter 11 was originally designed as a general protection for corporations against nationalization of industries as a measure to encourage more FDI, in the end it has had enormous significance for the environment. Of the twenty-seven suits filed under chapter 11 of NAFTA since 1994, eight are cases with clear environmental implications (see table 6.1). A number of these have been suits dealing with disputes over environmental regulations where firms have claimed that the imposition of certain environmental regulations violates chapter 11 in that new laws were deemed to have caused significant losses in profits for the complaining firm. In the eight environment-related cases, over US$1.5 billion in damages has been claimed by investors.

Social greens and bioenvironmentalists have lambasted chapter 11 of NAFTA. To them, it is a clear attempt to undermine environmental regulations and it encourages not just a regulatory chill but also a "race to the bottom." This is because they see the investment rules as restricting the ability of governments to protect the environment. For example, the suit by Ethyl Corporation against Canada in 1997 over its ban on imports of the chemical MMT (a gasoline additive) led the Canadian government to repeal its ban on that chemical. Moreover, the Canadian government had to pay US$13 million in damages to the corporation in

Table 6.1
NAFTA chapter 11 disputes related to the environment

Date suit filed	Company suing	Government being sued	Issue	Status
Oct. 2, 1996	Metalclad Corp.	Mexico	U.S. waste-management company challenges decision by Mexican local government to refuse it a permit to operate a hazardous-waste landfill in La Pedrera, San Luis Potosi, and decision by state government to create an ecological preserve in the area.	Tribunal ruled that Mexico violated Art. 1105 (min. standards of treatment) and 1110 (expropriation and compensation). Mexico was ordered to pay U.S. $16.7 million. Mexico applied for statutory review of the award before the Supreme Court on grounds that the tribunal had exceeded its jurisdiction. The court allowed most of the award to stand. The case was settled in Oct. 2000; Mexico paid undisclosed damages.
Dec. 10, 1996	Robert Azinian et al. (*Desona*)	Mexico	U.S. waste-management company challenges Mexican court ruling revoking its contract for nonperformance of waste disposal and management in Naucalpan de Juarez.	In Nov. 1999, the tribunal dismissed the investors' claims.

April 14, 1997	Ethyl Corp.	Canada	U.S. chemical company challenges Canadian ban on import and interprovincial trade in gasoline additive MMT, which automakers claim interferes with automobile onboard diagnostic systems. Manganese-based MMT is also a suspected neurotoxin.	After preliminary tribunal judgments, Canadian government repealed the MMT ban, issued an apology to the company, and settled out of court for US$13 million. (The interprovincial aspect of the trade ban had previously been found to violate Canada's nonbinding Agreement on Internal Trade.)
June 30, 1998	U.S. Waste Management Inc.	Mexico	U.S. waste-management company challenges state and local government actions in contract dispute with a Mexican subsidiary over waste-disposal services in Acapulco.	In June 2000 the Tribunal ruled that it lacked jurisdiction—Waste Management Inc. had not properly waived domestic legal claims as required by NAFTA. The investor resubmitted its notice of intent. The tribunal subsequently confirmed its jurisdiction; claim is proceeding.
July 22, 1998	S.D. Myers Inc.	Canada	U.S. waste-disposal firm challenges temporary Canadian ban (Nov. 1995 to Feb. 1997) on export of toxic PCB wastes.	Tribunal ruled that Canada violated art. 1102 (national treatment) and 1105 (min. standards of treatment). It awarded U.S. $5 million plus interest in damages. Canada applied to the Federal Court of Canada to set aside the award, but it was denied.

Table 6.1
(continued)

Date suit filed	Company suing	Government being sued	Issue	Status
Dec. 2, 1998	Sun Belt Water Inc.	Canada	U.S. water firm challenges B.C. water protection legislation and moratorium on exports of bulk water from the province.	Canadian government asserts that the claim is inactive, while the investor asserts that it is still pending.
June 15, 1999	Methanex Corp.	United States	Canadian chemical company challenges state phaseout of MTBE, a gasoline additive that has contaminated ground and surface water throughout California.	Tribunal has accepted admissibility, asked Methanex for more evidence supporting the allegation that California Governor Davis was improperly influenced by a competitor when he ordered the ban.
Nov. 6, 2001	Crompton Corp.	Canada	U.S. chemical company challenges Canadian ban on use of lindane, a known carcinogen, in canola seeds and seed treatments.	Canadian Tribunal process pending.

Source: WTO Watch website: www.tradeobservatory.org.

an out-of-court settlement. In effect, critics say, with chapter 11 the way it is, governments will be afraid to implement new laws for fear of being sued.[110] For this reason, they would like to see the investment provisions removed or rewritten. Institutionalists and market liberals have tended not to see these cases as ones that threaten environmental regulations in a blanket fashion. Rather, they argue that the tribunal in each case had solid and specific reasons for each decision, and maintain that there is no bias in chapter 11 against environmental regulations.[111]

Another set of investment rules through the OECD, the Multilateral Agreement on Investment (MAI), was negotiated in the mid to late 1990s. This agreement planned to enshrine NAFTA chapter 11–type rules, in particular the investor rights' provisions, on a much larger scale. There was an attempt at the time the Uruguay Round was negotiated to include such measures in an agreement under the WTO, but many developing countries resisted this. For this reason, the OECD sought to develop its own agreement, with a view to eventually including it in the WTO once it gained broader support.[112] The MAI fell through in the late 1990s, however, after a broad coalition of activists publicized the secretive nature of the talks and revealed the problems with the agreement, including its lack of environmental safeguards.[113] While OECD governments abandoned the MAI, many activists are still concerned that governments will slip such rules into other global agreements, because investment issues are under consideration in the Doha Round of trade negotiations at the WTO. The victory of activist groups in opposing this effort gave much steam to the antiglobalization movement.

Following the defeat of the MAI, the OECD revised its voluntary set of *Guidelines for Multinational Enterprises*. Originally adopted in 1976, these guidelines have periodically been reviewed and revised by the OECD, most recently in 2000. The 2000 revisions include recommendations that bring the guidelines into line with international environmental standards and treaties and with the Rio Principles.[114] For example, the newly revised guidelines promote already-existing environmental-management standards, such as the ISO 14000. But going further than ISO standards, they also call for TNCs to adopt "measurable objectives, and where appropriate, targets for improved environmental performance." The guidelines also call for better information on the environmental activities of firms, as well as more consultation with affected communities.[115] Critics, however, remain skeptical.

The OECD *Guidelines for Multinational Enterprises* only apply to the OECD. In 1977, the UN Center for Transnational Corporations (UNCTC) launched negotiations on a globally applicable voluntary code of conduct for TNCs, which included provisions on environmental conduct and outlined rights and responsibilities of TNCs.[116] The UNCTC, which was set up in the early 1970s, was mandated to monitor the economic, social, and environmental impact of TNCs, particularly those operating in developing countries. The code of conduct aimed to ensure that foreign direct investment did not have adverse consequences in these areas. Talks on this code were ongoing until the early 1990s, but it was never finalized or adopted. Under pressure from the United States, the UNCTC was dismantled just prior to the Rio Earth Summit, and the United Nations Commission on Trade and Development (UNCTAD) took over its remaining activities. Instead of implementing the UNCTC code, the Earth Summit promoted voluntary initiatives developed by corporate actors themselves, as discussed earlier.[117]

Social greens and bioenvironmentalists are critical of the current investment rules and instead call for harmonized standards globally, preferably ones that discourage unsustainable investment and, in the case of social greens, ones that promote local production. Greenpeace and Friends of the Earth, for example, have both advocated the adoption of a binding set of rules to ensure better performance of TNCs on environmental, social, and human rights. Their proposals call for firms to adhere to the highest standards, for corporate liability for damage, and for implementation of the precautionary principle.[118] (Chapter 8 discusses these proposals more fully.) Institutionalists would prefer an agreement that prevents a race to the bottom, but many do not support harmonizing standards. They do not want to discourage global investment (under the proper conditions), because it is crucial for sustainable development. While market liberals advocate a global investment agreement, they would like to see it focus on protection of corporate rights. Specific environmental-protection provisions in such an agreement, from their point of view, are not necessary.

Conclusion

Underlying assumptions about the nature of firms shape much of the debate over investment and the environment. Market liberals see firms as efficient managers, and as engines of economic growth. Most accept that negative environmental impacts of firms are a problem on occasion, but they nevertheless argue that the best approach is to let market incentives encourage firms to *voluntarily* work for better environmental management. Interfering with the natural pursuit of corporate profits is, as with global trade, unlikely to actually conserve nature. In fact in most cases it simply perverts ecological, social, and most significantly, financial goals. Institutionalists have sympathies with the market liberals, but they are more willing to see corporate activity as potentially negative more of the time. They argue that a strong state with a solid regulatory capacity and a vibrant civil society are necessary to reduce the impact of externalities and maintain a stable investment climate to ensure that firms have the proper incentives for sustainable development. Globalization, for many institutionalists, is also opening more possibilities for the global community to hold firms accountable, even those working in remote and weak states. This includes global agreements and standards to help evaluate corporate environmental performance. For institutionalists, these may be voluntary or binding.

Bioenvironmentalists and social greens have long condemned the ecological and labor practices of TNCs, particularly those operating in developing countries. In recent years globalization has bolstered the economic might of TNCs relative to small states and, not surprisingly, activists have begun to campaign even more vigorously against TNCs. The rather pejorative term *corporate globalization* signifies their concern that globalization is really about constructing a global political economy in the commercial interests of TNCs. Critics of corporations are suspicious of corporate motives as well as of their structural capacity to improve environmental performance. Both bioenvironmentalists and social greens call for strict controls on TNCs. They point to the structural factors within TNCs that drive firms to pollute the world's air and water and deplete the world's oil, minerals, timber, and fish, especially in the developing world. Weak states, corruption, and greed allow firms to get away with reckless environmental abuses. Relying on green markets and voluntary corporate initiatives might do some good—but ultimately, it is critical to simply prevent some of the corporate exploitation of the global environment. For bioenvironmentalists it is essential to control *both* TNCs *and* local firms—both are equally capable of plundering. Social greens take a slightly different tack. They call for measures to dismantle TNCs and replace them with smaller firms accountable to local people and local environments. This must occur alongside a broader reform of the global political economy, including, as the next chapter discusses, an overhaul of global financing.

7

Global Financing and the Environment

Global financing is increasingly at the heart of the economies of the developing world. This involves governments constantly borrowing and repaying public and private loans, a steady stream of multilateral and bilateral grants and technical assistance, as well as staggering private financial flows to corporations and individuals. Each of these has profound and somewhat different environmental implications. While the broad split in views again falls along the now-familiar coalitions between social greens and bioenvironmentalists on the one hand and the market liberals and institutionalists on the other, each perspective has its own unique take on financing and the environment.

Market liberals and institutionalists both see international financing in a broadly positive light. For market liberals it fosters global investment and economic growth, which benefits all countries by promoting global economic stability. It also specifically helps support efforts by poorer countries to develop, which enables them to better manage some of the inevitable environmental costs of early industrialization and natural resource exports. Market liberals, though, are wary of endorsing too much development aid, which they argue can interfere with market incentives. Institutionalists agree with market liberals on the benefits of global financing, but add that it is also one of the most vital financial and technical mechanisms to sustain environmental institutions and norms. They see financing as the primary means of building the institutional capacity of developing countries, and ultimately of the international community, to manage local and global environmental affairs.

Bioenvironmentalists and social greens are highly critical of current patterns of global financing. Both perceive it as yet another global driver of economic growth, and as a source of funds and technical advice that allow developing countries to overexploit natural resources, urbanize, and industrialize. Development assistance in particular wrongly assumes that elevating the living standards of poor countries to the level of the rich countries is a sustainable goal, while doing little to address the dangerous fact that the ecological footprint of the world's wealthy is already well beyond the earth's carrying capacity. Some extreme bioenvironmentalists add that development assistance is utterly misguided, merely compounding the ecological stress on the global ecosystem by supporting the overpopulation of the poor. To survive, in the words of Garrett Hardin, the rich must live "by the ethics of a lifeboat" and not perversely toss aid to the hordes who threaten to swamp the boat of wealth and prosperity.[1] Most social greens on the other hand see such extreme arguments as callous and immoral. Still, they agree that development financing is failing. It *does not* serve the interests of communities and individual well-being in the developing world, but instead serves the interests of international agencies, powerful states, and TNCs intent on exploiting the resources and peoples of the developing world. Far too much global financing supports industrialization and intensive agriculture in poor countries, and far too little supports healthy communities and sustainable livelihoods.

This chapter explores the environmental implications of some of the main channels of global financing. It focuses in particular on development assistance and developing-country debt, as well as assistance geared toward improving environmental management. It also covers the environmental implications of export credit agencies and private financial flows. It begins with an overview of various kinds of financing as well as recent trends in international finance.

Scope and Trends in International Finance

The globalization of financing is increasingly connecting the world in terms of money flows, tied to both trade and corporate investment. Financing at the international level occurs in a variety of ways. International development aid consists of grants (aid that does not have to be repaid), technical assistance (expert advice), and preferential loans (i.e., loans with lower interest rates and longer payback periods than standard public and bank loans). Multilateral aid is channeled through institutions like the World Bank and UN agencies. Bilateral aid is from one government to another. Multilateral and bilateral loans are referred to as *official debt*. Such loans may or may not be counted as official development assistance (ODA), which is assistance to multilateral institutions and developing-country governments that meets the guidelines of the twenty-two-member OECD Development Assistance Committee (DAC). To be counted as ODA, such assistance must be given explicitly for the promotion of economic development and welfare in developing countries, and it must also include a grant element of at least 25 percent. Much of the globe's international development assistance is ODA, but much of the world's official debt arises, too, from loans borrowed on terms just outside of this definition (although these terms are generally better than a private bank loan). Governments and private interests also borrow directly from commercial banks (referred to as *commercial debt*). Global financing comprises other private financial flows as well, such as markets for securities (stocks and bonds), and foreign-exchange trading.

Development aid—that is, grants and the subsidized portion of loans— is an important source of financing for poor countries. In 1970 the UN adopted a goal that donors should contribute 0.7 percent of their gross national product to developing countries. This was reiterated in the call for a New International Economic Order (NIEO) in 1974, as well as at the Rio Earth Summit in 1992 and again in recent years as part of the UN's Millennium Development Goals. This goal was reinforced in 2002 at both the Monterrey Financing for Development Conference and the Johannesburg World Summit on Sustainable Development. Yet throughout this time donor governments have fallen far short of this goal. Only five countries have reached or exceeded it, and the average for the OECD now stands at just 0.22 percent of gross national income of the DAC (figures 7.1 and 7.2). This percentage has been gradually declining since the early 1960s when it stood as high as 0.5 percent. In absolute terms, however, the total amount of money flowing into the developing world in the form of aid rose over that same period, despite dropping in the early 1990s and remaining stagnant since then. In 2001, ODA from DAC countries amounted to US$52.7 billion (figure 7.3). The United States contributed a mere 0.11 percent of gross national income, the lowest of the major donors (figure 7.2).

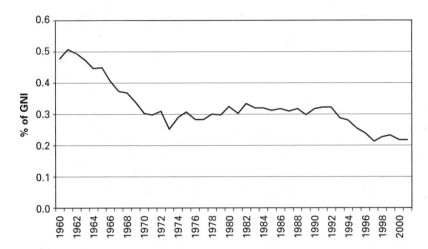

Figure 7.1
OECD DAC: Official development assistance: ODA as percentage of OECD GNI. *Data source*: World Bank World Development Indicators: www.worldbank.org/data.

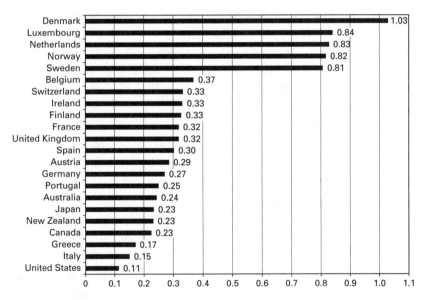

Figure 7.2
Official development assistance by DAC donor: ODA as percentage of GNI. *Data source*: World Bank World Development Indicators: www.worldbank.org/data.

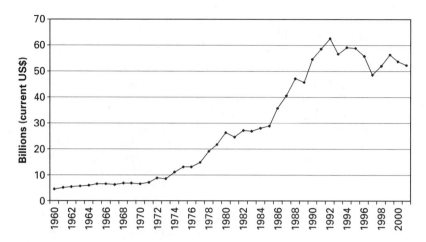

Figure 7.3
OECD DAC: Official development assistance: Total. *Data source*: World Bank
World Development Indicators: www.worldbank.org/data.

As official aid as a percentage of donors' income has been falling,
international debt has been rising, especially for developing countries.
External debt has become a defining feature of most developing-country
economies over the past 30 years (see box 7.1). In 1970 the total debt
stocks of developing countries stood at US$72.8 billion. It was US$609.4
billion by 1980, US$1458.4 billion by 1990, and a staggering US$2442.1
billion in 2001 (figure 7.4). Governments must make regular payments
on both the principal and the interest on this debt. These payments,
referred to as *debt service payments*, are a significant drain on the
economies of most developing countries. Figure 7.5 shows that the debt
service for developing countries has averaged between 15 and 20 percent
of their exports of goods and services (this is referred to as the *debt
service ratio*) over the past few decades. But there are wide variations.
The debt service ratio as a percentage of exports of goods and services
for Somalia in 1986, for example, was 66 percent, while for Belize in
that same year it was 10 percent.[2] In 2001 total debt service paid
by developing countries was US$377.5 billion (figure 7.6), of which
US$261.3 billion was paid in principal repayments and US$116.2 billion
was paid as interest repayments.[3]

The liberalization of controls on financial flows over the last two
decades has expanded global private financial flows well beyond

Box 7.1
The developing-country debt crisis

The ballooning of developing-country debt over the past 30 years is the product of a number of factors, both within developing countries and internationally. In the early 1970s, as a result of an excess of dollars in international markets linked to the OPEC oil-price rises and rising inflation, developing nations were encouraged to borrow at flexible-interest rates with few conditions from banks and governments. High inflation meant that real interest rates (the rate of interest adjusted for inflation) were quite low, making borrowing appear to be cost free for many countries. Commercial debt—that is, loans from private banking institutions—grew mainly in Latin America and Asia. Official debt—in other words, loans from governments and multilateral institutions—affected all parts of the developing world and formed the main source of loans for sub-Saharan Africa. Lenders at this time made the assumption that countries would not default and were aggressive in offering them loans.

After these debts began to accumulate, however, changes in the global economy made the loans increasingly difficult for poor countries to repay. In 1979, OPEC nations doubled the price of oil, and in response in 1981 the U.S. Federal Reserve raised U.S. interest rates above 20 percent to fight inflation. International interest rates rose as well, sparking a worldwide recession. Demand for the products sold internationally by developing countries plummeted, and the prices of raw commodities, which made up a significant proportion of developing-country exports, did not rise in step with the prices of the industrial products they were importing. The result was a double blow for developing countries with commercial debt. They were faced with higher real interest rates (because the rate of inflation fell as interest rates rose) and, because of the recession, they were not earning enough income to repay the loans. Countries holding official debt were also affected, because the global recession made it extremely difficult for them to earn enough from their exports to repay their loans. In 1982 Mexico announced it was unable to pay interest on its foreign debt. Other countries followed and by the mid-1980s the world was immersed in a full-blown debt crisis. In return for loan refinancing by the IMF and the World Bank, indebted countries agreed to adopt strict structural adjustment policies dictated by these institutions (discussed later in the chapter).

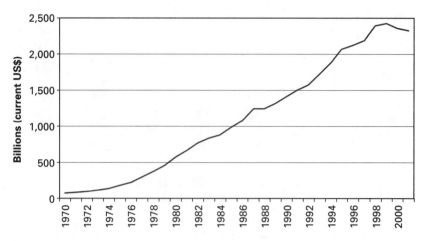

Figure 7.4
Debt owed by developing nations. *Data source*: World Bank World Development Indicators: www.worldbank.org/data. *Note*: Total external debt is the sum of public, publicly guaranteed, and private nonguaranteed long-term debt, use of IMF credit, and short-term debt.

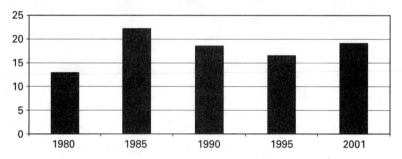

Figure 7.5
Developing nations: Debt service ratios (percent of exports of goods and services). *Data source*: World Bank World Development Indicators: www.worldbank.org/data.

international bank lending. Over a trillion dollars in national currencies is now traded across borders every day. As shown in figure 7.7, in April 1989, total average global foreign-exchange market turnover was US$590 billion per day. This figure had more than doubled to US$1.2 trillion by April 2001 (despite a fall of 19 percent between 1998 and 2001).[4] Foreign speculation now far outweighs trade in goods and services, with the value of about 10 days worth of transactions on the world

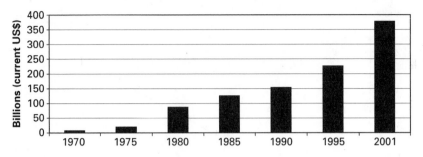

Figure 7.6
Developing nations: Total debt service paid. *Data source*: World Bank World
Development Indicators: www.worldbank.org/data.

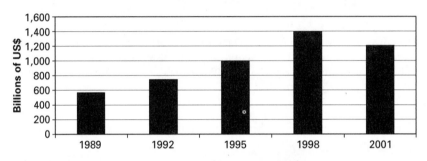

Figure 7.7
Global foreign-exchange market turnover. *Data source*: Bank for International
Settlements press release 31/2001E, 2001.

financial markets approximately equal to the value of annual world
production of goods and services. Trade in stocks and bonds has also
ballooned in the past decade, with new stock exchanges opening in over
seventy countries in that period.[5]

Multilateral Lending: The World Bank and the IMF

Multilateral lending is a defining feature of international finance and the
environment. The World Bank, as the main multilateral lending agency,
is of particular significance. The World Bank not only plays a crucial role
in lending to developing countries, but also in establishing the adjustment programs designed to ensure repayments of debts—both commercial and official. The World Bank provides a significant proportion of all

multilateral assistance to developing countries, and its policies are extremely influential over the policies of other development banks and donor states.[6] This influence affects some regions more than others, and is especially important in areas like sub-Saharan Africa, where multilateral aid comprises the bulk of foreign assistance.

The World Bank and the International Monetary Fund (IMF) were established in 1944 after delegates from forty-five countries negotiated a monetary agreement at the UN Monetary and Financial Conference in Bretton Woods, New Hampshire. The Bretton Woods Agreement— which established a system of convertible currencies, fixed exchange rates, and free trade—was designed to support a liberal international economic order. The IMF was to be the overseer of exchange rates and provider of funds to meet balance-of-payments deficits. The original purpose of the World Bank, officially known as the International Bank for Reconstruction and Development (IBRD), was to finance projects to reconstruct a devastated Europe after World War II. The original task of the World Bank was largely complete by the late 1950s and early 1960s. By then there were a host of newly independent countries on the world stage. The World Bank therefore turned its attention to granting loans to the developing world, partly to stimulate global economic growth, and partly to ensure global economic stability. In 1960 the World Bank established a separate lending arm, the International Development Association (IDA), which specifically lends to the poorest developing countries on "soft" terms (low interest rates).[7]

Today the World Bank is the largest development lending agency in the world, lending approximately US$19.5 billion to developing countries in 2002.[8] The World Bank lends to support both specific projects and for structural adjustment. It disburses lending for structural adjustment more quickly, supporting shortfalls in balance of payments in return for policy changes in the borrowing country. The World Bank introduced structural adjustment lending in 1980, as a response to the economic stagnation of poor countries with high debt loads. This type of lending has become a significant feature of the institution's lending portfolio since then, with a climbing percentage of World Bank loans going to support adjustment programs. In the 1980s adjustment lending averaged around 17 percent of its loans. In 2001 33 percent of its loans were for adjustment, with around two-thirds supporting projects.[9]

A market-liberal framework shaped the views within the World Bank on the environmental impact of its lending for most of the postwar period. Over the last few decades, however, the Bank has slowly ebbed into a more institutionalist perspective, advocating, for example, global institutions for environmental protection.[10] Indeed, the World Bank was the first multilateral development agency to establish an Office of Environmental and Health Affairs (later renamed the Office of Environmental Affairs) as far back as the early 1970s,[11] although it did not begin to seriously address the environmental impacts of its loans until the 1980s and 1990s, in the face of mounting criticism from environmentalists.

Critique of Project Lending and the World Bank's Response

Environmental activists from around the world focused their campaigns on the World Bank's megainfrastructure projects and large-scale migration and resettlement schemes.[12] Critics, mainly social greens and bioenvironmentalists, attacked the Bank's market-liberal assumption that economic growth would cure environmental problems. At the time the Bank, like most other international lenders, paid little attention to environmental issues in project design. This was not surprising, because most lenders, including the World Bank, did not have the staff to analyze every project to assess environmental impacts, let alone to incorporate environmental goals into the design of every project. The Office of the Environment had few staff members and was separate from the project-design office. In large part this neglect was natural, because Bank officials were mostly traditional liberal economists who worked on the assumption that good economic policy was good for the environment.[13] Any environmental problems in the wake of projects were seen to be negative externalities inherent in the development process.[14]

The Bank's large infrastructure projects—such as roads, dams, and power plants, as well as industrial agricultural projects and migration schemes—did not account for environmental concerns. Many projects, as a result, contributed directly to soil erosion, increased rates of deforestation, a loss of native genetic material for agriculture, and the disruption of the lives of indigenous peoples. Even where the World Bank did have environmental policies in place in the 1980s, these policies were generally ignored.[15] Environmental groups were able to track the environmental performance of a number of World Bank projects, assisted by a growing network of local groups on the ground. Two World Bank projects drew particular media attention: the Polonoroeste project in the Amazon and the Narmada Dam Scheme in India.

The Polonoroeste ("northwest pole") project was a colonization scheme that aimed to move poor settlers from urban areas in the south of Brazil to the northwest region in the Amazon forest. The World Bank lent US$457 million to Brazil for this project in the early 1980s, the only non-Brazilian source of funds.[16] To facilitate the colonization of the Amazon, a major highway was built into Rondônia state, known as BR-364. Half of the World Bank's loans for this project went toward financing the construction of the highway and feeder roads. The World Bank, for its part, claimed it would ensure that the project was environmentally and socially sustainable. The critics felt this was simply impossible. In the first half of the 1980s nearly a half million settlers flooded into Rondônia. The government gave colonists a plot of land to grow food, often land already occupied by native peoples. Huge swaths of forest were cut down and burned to build the highway and clear new land for the colonists (with the aim of growing export crops). The soil was not suitable for agriculture, however, and the farms failed. Much of the deforested land was then used for cattle ranching, which failed too. This project brought terrible social hardship. Many of the native peoples of Rondônia did not survive the new diseases brought in by the settlers. This project also left a huge scar in the Brazilian rainforest. The percent of Rondônia that was deforested increased from 1.7 percent in 1978 to 16.1 percent in 1991.[17]

The Narmada Valley Projects in India is a development scheme whose goal is to establish a series of dams along the Narmada River for irrigation and power generation. The Indian government conceived of the project, then approached the World Bank for financial assistance. The Bank pledged US$450 million in 1985.[18] The Indian government needed to resettle thousands of people to build the dams. The World Bank's own policy on resettlement at the time required that the Indian government have a resettlement and economic rehabilitation plan that was acceptable to the displaced peoples in place before the loan was approved, but this did not occur.[19] Criticism of the dams mounted in the late 1980s and early 1990s, and the Narmada Bachao Andolan (Save the Narmada Movement) was formed in India, with transnational support, to protest the dams.[20] Activists highlighted the environmental and social harm from displacing tens of thousands of people, including the impact on marginal lands, deforestation, and community well-being. The dams themselves, activists argued, would irrevocably disrupt the ecosystems and watersheds.[21]

Armed with detailed information about environmental abuses in these and a growing number of other projects, environmental NGOs launched a concerted and sustained campaign to "green" the World Bank. Environmental groups had long tried to educate the public about the poor environmental practices at the World Bank. But their strategy took on a new angle in the early 1980s, as they began to lobby donor-government legislators responsible for approving government donations to the IDA of the World Bank. The U.S. government in particular was targeted, because it has traditionally been skeptical of multilateral organizations (and thus frequently withholds funds from them). Environmental groups—such as the Environmental Defense Fund (now Environmental Defense)—testified before the U.S. Congress on the negative environmental impact of the Bank's various projects, most notably the Polonoroeste project.[22] This prompted the United States to question the World Bank's environmental record. The United States did agree in the end to give its donation to the IDA replenishment—but with qualifications for the Bank to improve its environmental practices.[23] This led to a wider NGO campaign in North America, Europe, and the Pacific to reform the World Bank's environmental policies.[24]

The World Bank halted the Polonoroeste project in 1985—the first time an ongoing project loan was halted on environmental grounds.[25] Environmental groups were able to pressure the World Bank to incorporate environmental issues more broadly in its project lending. The World Bank president at the time, Barber Conable, admitted in 1987 that the Bank had made mistakes in the past when it came to the environment and the displacement of people, and he promised to make changes to make the bank more "green."[26] The Bank conducted an independent review of the Narmada Valley projects in the early 1990s. The results, released in 1992, highlighted the severe problems associated with the project.[27] A year later the Indian government and the World Bank decided to discontinue the loan for the Narmada Dam scheme, opting for Indian self-funding rather than face continuous international environmental opposition.[28]

The World Bank undertook a major restructuring in 1987, which included the reorganization of the environment staff from the Office of the Environment to a new Environment Department as well as into Regional Environment Divisions.[29] The Office of the Environment at the time had just five specialists.[30] The new Environment Department was charged with researching the relationship between the environment and development, as well as assessing projects with potential environmental implications. After 1991 the World Bank also began to require environmental-impact assessments of all projects. By 1990 the Bank had increased the environment staff by about 60 people; by 1994 nearly 200 had been added.[31] The Bank in addition pledged to design more projects with positive environmental benefits, and to consult with affected peoples and NGOs before embarking on projects with the potential to cause environmental harm.

The World Bank also helped to set up a pilot project for the Global Environment Facility (GEF) in 1991. The purpose of the Facility, corun by the World Bank, the UN Environment Programme (UNEP), and the UN Development Programme (UNDP), is to finance projects with global environmental benefits in developing countries with a per capita income of US$4,000 or less (we discuss the GEF more fully later). In 1992, to coincide with the Rio Earth Summit and to advertise its new green credentials, the Bank devoted the *World Development Report* to the theme of Development and the Environment.[32] This initial process of greening continued into the mid-1990s.[33] In 1993 the World Bank established an independent Inspection Panel to review claims by peoples affected by World Bank projects—a body seen by some observers as "remarkably" autonomous.[34]

Collectively these changes amount to a significant shift in World Bank policies with respect to the environment in a relatively short period. The World Bank, increasingly drawing on the language of institutionalists, claimed it was "mainstreaming the environment" into its everyday practices. This meant far more funds for environmental projects.[35] Not everyone is convinced of the Bank's newfound environmental awareness and responsibility. Critics accuse the Bank of "greenspeak." No doubt, these critics note, there is a new *rhetoric* and many new *reports* on the environmental activities of the Bank. Yet environment personnel at the Bank are still marginal players,[36] and in practice most decisions on whether to lend funds for projects are made independently of the environment department. The most important criterion remains the economic rate of return. Projects with environmental benefits, critics note, tend to be small-scale and are of little interest to the Bank lenders because of the time required to design and manage them. Such projects are sidelined in part because the lending culture at the Bank tends to reward employees who lend large amounts, and not on the outcome of the loan itself (which is far harder to measure). The culture thus favors larger projects, the very ones with the greatest potential to cause environmental harm.[37] Critics note further that the concern with environmental matters in the first half of the 1990s seemed to fade in the latter half of the decade. In 1994 environmental lending made up 3.6 percent of total lending by the World Bank; loans for projects with environmental benefits increased some thirtyfold from 1990 to 1995. These upward trends did not continue, however, and environmental lending had fallen to just 1.02 percent of total lending by 1998.[38]

Critics also complain that the Bank does not adequately consult NGOs regarding the potential environmental impacts of project loans. NGOs are now calling for more far-reaching reforms both to the consultation process as well as to environmental lending. A Friends of the Earth report notes that "in fiscal year 2000, close to half the lending from the World Bank's private sector divisions was [still] in environmentally harmful sectors, such as oil, gas, coal, mining, chemicals, and infrastructure projects."[39] Friends of the Earth is calling on the Bank to adopt an exclusion list that would prevent it from lending for projects in sectors that cause severe environmental harm. The list includes, for example, projects for infrastructure or resource extraction in frontier or primary forests as well as projects in areas the UN has designated as ecologically significant. It includes, too, projects that involuntarily resettle more than 500 people, large dam projects that do not meet the criteria set out by the World Commission on Dams, and any logging or extractive project without an accredited NGO certificate of sustainability.[40] NGOs are also pressing for more institutional accountability. The process of opening up of the Bank has, in the view of many NGOs, only just begun. Thus far, for example, there is little evidence of internalization of environmental costs, and environmental issues are not yet "fully integrated into the core logic of the World Bank's development strategy."[41] An internal World Bank evaluation affirms the claims of many environmental NGOs that the quality of environmental assessments deteriorated in the late 1990s.[42]

The World Bank, after a two-year consultation process, launched a new environmental strategy in July 2001. It focuses on three core objectives: "improving the quality of life; improving the quality of growth; and protecting the quality of the regional and global commons."[43] Financial and technical support for sustainable development—understood as economic development, social cohesion, and environmental protection—is a core component of the World Bank's mandate of alleviating poverty.[44] It is too early to evaluate the impact of this new environmental strategy. But, predictably perhaps, the critics see it as just another coat of gloss over the fundamental truth that the Bank is just that: a bank.

Structural Adjustment Lending and the Environment

One-third of World Bank lending currently supports structural adjustment programs (SAPs). This type of loan (structural adjustment loan, or SAL) provides balance-of-payments support to indebted governments. In return for these funds, governments agree to reform economic policies following the advice of the World Bank and the IMF. The intended purpose of this economic restructuring is to promote growth and eventually enable indebted governments to repay their debts. The World Bank began to undertake this type of lending in the early 1980s after many developing-country governments began to experience difficulties repaying external debts. These structural adjustment loans have enabled the IMF and World Bank to wield considerable influence over economic policies and outcomes in poor regions of the world. Not only are these agencies themselves a significant source of funds to developing countries, but their policy advice also influences other development banks as well as donor states. Sub-Saharan African and many Latin American countries have been most profoundly affected by structural adjustment lending. The policy reforms called for under such loans have considerable environmental implications. Do these reforms help or harm the environment?

Not surprisingly, market liberals and institutionalists perceive the reforms under structural adjustment programs as critical to sustainable development. The main types of policy changes called for are those that open up developing-country economies to the world economy. The purpose is to facilitate trade and investment and ultimately economic growth. The IMF and World Bank had long been critical of any measures that cut off developing economies from world markets, such as interventions in markets for commodities and currencies. Along these lines, the first condition on SALs is usually currency devaluation and the implementation of a floating exchange-rate system. The intent of devaluing the local currency is to boost the income of local producers and exporters (assuming at least the same amount is traded). This has several further benefits for the local economy. It can lead to increases in production, which can lower prices and raise the volume of exports, allowing developing countries to recapture lost markets.

Other reforms called for in adjustment programs include a liberalization of domestic price policies and a removal of restrictions on trade. These measures would help to "get the prices" right. The World Bank sees excessive government intervention as the primary explanation for why actual prices are lower than market prices. The SAL conditions also call for the privatization of state-owned firms, reductions in government spending, and a relaxation of restrictions on foreign investment, all designed to reduce inefficiencies. Together, the adjustment reforms should make it easier for a country to repay debts because they encourage activities that earn foreign exchange and reduce government inefficiencies. SALs were supposed to last 3–5 years only, and were seen as a one-time adjustment. In fact, SAPs have been in place in a number of countries for 20 years. Most countries in sub-Saharan Africa, and many in Latin America as well as a number in Southeast Asia, had adopted SAPs by the 1990s. These reform programs have become a dominant feature of their economies.

The IMF and World Bank did not at first consider the environmental dimensions of SAPs. This is understandable since, as noted, the Bank did not put a large amount of effort into assessing the environmental impact of its lending in the early 1980s.[45] But it also did not think that such scrutiny was warranted, because it was assumed that the economic growth stimulated by market-oriented reforms was necessary to improve environmental management anyway. This line of reasoning has continued to characterize the World Bank's response to criticism of the ecological implications of SAPs. The Bank has argued that the impact of SAPs on the natural environment is hard to generalize, but that in most cases it has probably been either neutral or positive.[46] A World Bank report notes that "65 percent of recent SALs included an explicit environmental section, generally consisting of a statement to the effect that no environmental effects could be expected."[47]

Market liberals at the World Bank see the potential for some direct ecological benefits of SAPs. They can promote conservation by raising prices of previously underpriced natural resources, such as timber and minerals, to levels closer to true market value.[48] Price liberalization and currency devaluation also encourage the production of export crops like coffee, rubber, palm oil, and cocoa. The root systems of these tree crops can help prevent soil erosion. The removal of government subsidies, for example on fuel, agricultural chemicals, and industries, can improve environmental performance because these subsidies support inefficient industries and practices. Removing subsidies should reduce inefficient consumption—such as single-occupancy vehicles—which means less pollution and less chemical use.[49] And because SAPs encourage private ownership of both businesses and land, it creates, in the World Bank's view, incentives for stewardship and conservation.[50] If there are unexpected adverse effects from adjustment programs, these generally arise from imperfect policies, markets, or institutions.[51]

Critics, social greens in particular, have attacked SAPs as significant sources of unsustainable development in adjusting countries. Many claim that poverty and inequality have increased under adjustment programs, in clear contrast to World Bank declarations. Higher rates of poverty only fuel environmental degradation, as people deplete natural resources to survive (although, as chapter 4 discusses, the poor are not to blame).[52] More directly, critics have attacked SAPs for increasing unsustainable exports of natural resources and encouraging pollution-intensive foreign investment. Many claim that SAPs virtually ensure that heavily indebted countries will turn to unsustainable natural resource exports in order to earn the foreign exchange to pay off debts.[53]

Some critics, for example, claim that currency devaluations and the removal of trade restrictions create strong incentives to export large quantities of timber, which in turn drives deforestation. Data show that the highest rates of deforestation are in the most highly indebted countries. It is difficult to make the direct link between SAPs and deforestation, but there are, critics assert, some striking patterns. Brazil is one of the most highly indebted countries, and it has one of the most alarming rates of deforestation. The rate of deforestation began to rise at the time of the debt-crisis outbreak and during the implementation of adjustment policies in the 1980s.[54] In Ghana deforestation rose dramatically after it adopted a SAP in the 1980s. Between 1983 and 1986, Ghana's exports of tropical timber rose by 42 percent and the volume of timber production tripled.[55] Commercial logging has contributed to the shrinking of the country's forest area to one-quarter of its original size. Critics also point to unsustainable increases in tropical timber exports from Cameroon, Côte D'Ivoire, the Philippines, Tanzania, Thailand, and Zambia following adoption of adjustment reforms.[56]

Export-oriented adjustment policies have also pushed countries to increase their agricultural exports, often requiring a change in crops and agricultural techniques. In many cases, but not all, this has led to a decrease in food production and an increase in export-oriented crops. In some cases, plantation owners and farmers have brought large areas of marginal land under cultivation, contributing to deforestation, flooding, high rates of soil erosion, and falling agricultural yields. One example is Senegal, where policies to encourage an increase in the production of groundnuts contributed to severe soil degradation and lower productivity.[57] Industrial agricultural methods are on the rise in many countries in a bid to increase exports in developing countries. Production of crops like cotton, flowers, and tobacco—with particularly harmful ecological impacts because of the chemicals and fertilizers needed to maximize productivity—often rises under SAPs because these crops generate foreign exchange (in some cases the World Bank expressly encourages the development of such crops). Production of all three of these crops, for example, increased in Tanzania under its adjustment program.[58]

Critics also blame SAPs for an increased interest in mining and mineral exports in the developing world. In the Philippines, the adjustment program led the government to revise the mining code to encourage more foreign investment in mining.[59] Critics claim, too, that the expenditure reductions called for under SAPs have a direct environmental impact in adjusting countries. Governments, to decrease spending, have cut many programs, including those geared to protect the environment. Many have axed funding for national parks and agencies to monitor enforcement of environmental regulations. Thailand's budget for pollution control, for example, was cut under its SAP by 80 percent between 1997 and 1999.[60]

The debate over structural adjustment lending and the environment is still ongoing. The World Bank, although it continues to see adjustment as an overall positive force for the environment, has begun to engage in the debate with its critics and has undertaken studies on ways to better understand the linkages. The IMF, on the other hand, has remained silent on the issue. Studies on both sides of the debate are often contradictory, and clearly there is a need for more research into the impact of SAPs on the environment.

Multilateral Environmental Aid and the GEF

Consistent with the 1987 Brundtland Report and the message put forth at the 1992 Earth Summit in Rio de Janeiro, institutionalists continue to stress that much more aid from rich to poor countries is necessary to help finance environmental protection in developing countries. Such financing, institutionalists assert, is best directed toward specific environmental activities within those countries, as well as toward global institutions that promote sound ecological practices.[61] To achieve this it is essential to ensure that more of the rich countries live up to their commitments. Far too often, institutionalists lament, this does not occur. At the 1992 Earth Summit, for example, rich countries were only willing to commit to providing US$125 billion of the necessary US$625 billion required to implement the conference action program to promote sustainable development (Agenda 21).[62] The total of all global assistance to developing countries stood at US$52.3 billion in 2001, far short of the total amount required to meet the goals of the Earth Summit. Both rich and poor countries agree that more aid is needed, but they disagree on where it should come from, how much is needed, and how it should be allocated. Furthermore, disputes have arisen over the "conditionality" of environmental aid—that is, over the restrictions donors attach to the aid, sometimes broad policy reforms as in the case of structural adjustment loans, or sometimes specific measures to protect the environment.[63] There have also been concerns raised about the "additionality" of aid, referring to whether aid for the environment is to be "new" aid that is over and above existing aid flows, or just refocused aid.

For social greens, the problem is less about the *amount* of aid and more about the *source* and *focus* of aid. It is essential in their view to restructure the global political economy to rectify global economic inequalities. This will require far more than just lump sums of aid and loans, but new trade, investment, and financial flows. It is critical to restructure the financial system so it empowers local agencies, rather than global agencies, with the funds to support sustainability.[64] Social greens tend to be suspicious of global institutions responsible for funding sustainability, because too often these institutions function with hidden agendas that are more about the interest of capitalists and donors than about the interest of communities of peoples.

This debate between institutionalists and social greens has played out over the environmental funding by the Global Environment Facility. The GEF undertaking was very much an institutionalist project, because it foresaw that global cooperation and funding through such an institution was essential for long-term environmental protection. From its inception, it has been a highly politicized institution. The GEF was set up with a budget of about US$1.3 billion just prior to the 1992 Earth Summit in Rio de Janeiro at the urging of several EU countries.[65] The initial idea was to create a global "green fund" to help finance sustainable development initiatives called for in Agenda 21. The impetus for the GEF was to provide funding in the form of grants for developing countries to help them pay the "incremental costs" of global environmental obligations under international environmental agreements. Incremental costs, difficult even for the GEF to define precisely, refer in a broad sense to the extra costs developing countries pay for projects with "global" benefits.[66] Along these lines, the GEF funds only projects with "global" environmental benefits. Initially these were restricted to projects that focused on international waters, climate change, ozone depletion, and biodiversity. More recently, the GEF has added land degradation and persistent organic pollutants to this list.[67]

The GEF is run by the World Bank, the UNEP, and the UNDP, each of which takes a role that mirrors its own expertise. The World Bank is the lead agency and handles the finances. The UNDP directs the technical assistance associated with GEF projects, and the UNEP coordinates between the GEF and the global environmental-agreement process.[68] It was foreseen from the beginning that NGOs would have a role in GEF activities, though they had no formal rights in the pilot phase of the project.[69] Funding for GEF projects comes from the GEF Trust Fund, a fund to which countries pledge donations. By 2003, the GEF had committed more than US$4 billion to over 1,300 projects in over 140 countries. In 2002 donor countries had pledged an additional US$3 billion to the GEF Trust Fund to cover project costs through 2006.[70]

During the early years the GEF was subject to heavy criticism from both environmental NGOs and developing countries. The World Bank quickly came to dominate the new institution, and there was in fact little consultation with NGOs and local communities on project design and implementation. Developing countries were unhappy with the GEF's commitment to only fund the "incremental" costs of projects. There was also concern that the dominant role of the World Bank gave too much weight to the needs of donors over the recipients. Attorney Bruce Rich at the U.S. advocacy group Environmental Defense succinctly explains: "The formulation of the Global Environment Facility was a model of the Bank's preferred way of doing business: Top-down, secretive, with a basic contempt for public participation, access to information, involvement of democratically elected legislatures, and informed discussion of alternatives."[71] For these reasons many NGOs were wary of endorsing the GEF as the one and only fund connected to global environmental treaties.[72]

Discussions on how to restructure the GEF to ameliorate these problems began as early as the 1992 Rio Earth Summit and continued over the next few years. Environmental NGOs and developing countries argued for a more accountable and democratic institution.[73] They were in favor of a voting structure similar to the UN system of one country one vote. They also argued for the management of the GEF to become independent from the World Bank. Environmental NGOs called for greater participation of nonstate actors, including NGOs and local communities, in GEF decision-making procedures and implementation. The donor countries on the other hand were most concerned about ensuring an efficient and effective institution. For this reason they preferred the World Bank to continue to manage the GEF, as well as a voting system similar to the Bretton Woods system of one dollar one vote.[74]

The negotiations ended in a compromise. The reformed GEF incorporates a mixed voting system with elements of both the UN and Bretton Woods voting structures.[75] The management of the GEF is formally housed in the World Bank, but is now functionally independent. The

GEF now consults more formally with NGOs on its projects, and NGOs are allowed to attend GEF Council meetings as observers and in some cases as participants, which is unusual in global institutions.[76] The GEF Council is made up of eighteen recipient-country governments and fourteen donor-country participants and five NGOs. The GEF was formally adopted as a permanent body in 1994.

Social greens remain skeptical of the GEF even though it is now undeniably a bit more democratic, transparent, and accountable. There are broad concerns about the "bandage" approach of an institution like the GEF. It papers over a few particular problems with money, rather than addressing the underlying structural and systemic causes. The GEF largely ignores, for example, issues like consumption.[77] Social greens have several specific concerns as well. GEF money is given in the form of grants. This is appropriate, indeed essential, social greens agree. Yet, because the GEF only pays for the "environment component" and only the part with "global benefits," most of the project grants are in fact in some way tied to other World Bank loans. This arises almost inevitably, critics argue, from the Bank's continued dominance of the GEF process. A report sponsored by the NGO Environmental Defense and The Halifax Initiative notes that the World Bank "has used the GEF to externalize environmental costs and increase the indebtedness of Southern countries by 'sweetening' proposed loans with green grants."[78] Social greens further accuse the GEF of focusing mainly on transferring technology to developing countries. Projects are, moreover, top-down and do not adequately incorporate the views and participation of the local people most affected by them.[79] It is especially difficult for local groups in developing countries, for example, to learn the language in which proposals have to be written to be accepted by the Bank.

Bilateral Finance: ODA and Export Credit Agencies

Official Development Assistance
Two-thirds of total concessional assistance to developing countries is in the form of grants.[80] Multilateral development banks mainly provide loans. The UN, Europe, and bilateral donors, however, primarily supply grants. Around three-quarters of the total concessional assistance of bilateral donors is in the form of grants. These grants support a broad

range of activities, including sustainable development projects, education programs, technology transfers, food relief, and emergency assistance. The motives behind bilateral aid arise out of complex historical, cultural, political, and increasingly commercial foreign-policy goals. Bilateral donors put less emphasis than multilateral development banks on the economic "performance" of the assistance and more on humanitarian assistance and comprehensive security concerns. Many bilateral donors ensure that well over 90 percent of their assistance is in the form of grants. In 1998 and 1999 the grant share of total ODA commitments for countries like Australia, Canada, Denmark, Switzerland, and Sweden was 100 percent. The U.S. grant share was 97 percent in 1998 and 98 percent in 1999. The UK grant share was 94 percent in 1998 and 92 percent in 1999. The great exception in terms of the grant share of ODA is Japan, the world's largest bilateral ODA donor in the 1990s. Japan's grant share was 35 percent in 1998 and 39 percent in 1999.[81] Japan is an interesting case because its philosophy of ODA, embedded in the logic of market liberals, challenges the more institutionalist approach of the other bilateral donors.

Japan became the largest bilateral donor in 1991. In FY1997 (April 1, 1997–March 31, 1998) Japan distributed over US$9 billion in ODA. Of this, about 20 percent was designated as environmental aid. Well over half of this US$9 billion, as noted, was in the form of concessional loans. Some argue that loans, even ones on highly concessional terms, are never "environmental" because the pressure of servicing and repaying foreign debt pushes a government to export larger and larger volumes of natural resources to earn foreign exchange. Yet the Japanese government has a reasoned defense of its focus on concessional lending even for environmental projects.

ODA must, in Japan's view, ultimately foster self-reliance and long-term incentives to move beyond aid. There must be incentives to ensure that elites in recipient countries draw on ODA in a way that sustains economic growth (and thus generates the income to repay in 25 years or so). The need to service and repay loans puts a much greater onus on recipients to ensure ODA actually "performs" in the long term. Grants are useful—indeed these are often essential for humanitarian and practical reasons—yet too much grant aid can create a culture of welfare and dependence, where elites in developing countries are content to rely on handouts to prop up state services and appear competent in the short term (and perhaps at the same time line their own pockets). Moreover, grants alone, the Japanese government argues, will simply not provide enough funds to build the infrastructure—water and sewage systems, roads, ports, hydroelectric dams, and airports—so critical for long-term economic growth. Donors must aim to graduate recipients from a dependence on aid, and the only way to do this is to ensure local economic growth. Most of Japan's concessional assistance has gone into countries in the Asia-Pacific region. The exceptional economic performance of this region after World War II confirms, in the view of the Japanese government, the long-term logic of a focus on loans and large-scale infrastructure. Japan's ODA has helped countries like Malaysia to move beyond a dependence on aid, and countries like China, Thailand, the Philippines, and Indonesia to move steadily in that direction. This success is unlike that in any other region of the developing world.[82]

Export Credit Agencies

Export credit agencies (ECAs) are another important source of bilateral finance. ECAs are public agencies in developed countries that provide credit—in the form of government-backed loans, investment guarantees, and risk insurance—to support both investment and trade in developing countries. These credits are specifically tied to business contracts with companies based in the lending country.[83] Developing countries that accept projects financed by ECAs are in effect taking out a loan from the ECA, which then goes to the corporation providing the service. The developing-country borrower (sometimes a government, sometimes a private entity) must eventually repay the loan, not the company involved from the industrialized country. Most ECA loans to private entities in the developing world require that their governments back the loans, often turning what appears to be privately held debt into public debt.[84] Most industrialized countries have at least one ECA that provides such assistance. Examples of some of the more prominent ECAs are the Export-Import Bank of the United States (the Ex-Im Bank), the U.S. Overseas Private Investment Corporation (OPIC), the Japan Bank for International Cooperation (JBIC)/International Financial Corporation (formerly JEXIM), the Canadian Export Development Corporation, the German Hermes Kreditversicherung-AG (Hermes), the UK Export Credit

Guarantee Department (ECGD), and the Australian Export Finance and Insurance Corporation (EFIC).[85]

In providing this financing as well as loan guarantees and insurance services, ECAs seek to assist corporations from their own country as well as to finance development projects in developing countries. These agencies provide a large and growing amount of development finance for trade and investment, at interest rates higher than those of other bilateral and multilateral assistance. The combined total of finance from all ECAs is now more than twice the total value of all development assistance, bilateral and multilateral.[86] ECAs provide some US$100–200 billion in loans to developing countries annually, and they subsidize nearly 10 percent of global exports, making them significant financial players.[87] Their importance for developing countries is stark. They account for some 40 percent of developing country debt owed to official agencies. Export credits, for example, account for 64 percent of Nigeria's external debt, and 42 percent of the Democratic Republic of Congo's external debt.[88]

ECAs have come under attack by environmental groups in recent years. The nature of export credit lending means ECAs tend to finance and insure ventures with high risks, including oil, gas, logging, and mining projects, as well as large-scale dams, nuclear power, chemical facilities, and road building in remote areas. As Aaron Goldzimer of the NGO Environmental Defense notes, "When an ECA will take on most of the risk and provide nearly full compensation if something goes wrong, there is every incentive for corporations and banks to move ahead with any overseas transactions—even excessively risky ones."[89] Critics charge that ECAs do not properly consider the environmental implications of these projects before funding them. ECA Watch, an international NGO campaign on export credit agencies, claims "half of all new greenhouse gas-emitting industrial projects in developing countries have some form of ECA support."[90]

Unlike bilateral ODA and multilateral development agencies, most ECAs are not subject to any environmental or social standards. They are also highly secretive agencies that do not disclose much information about the projects they finance.[91] The U.S. ECAs are an exception in this regard, because the United States is the only industrialized country to subject projects funded by its ECAs to environmental assessments. Under pressure by U.S. environmental groups, there has been growing attention in recent years to the environmental impacts of ECA loans. The case for more transparency has been particularly strong in the United States, because the U.S. ECAs have links to the U.S. Agency for International Development (USAID), which on paper has strict environmental standards.[92] This type of pressure is lacking in many other industrialized countries. The result is that when a U.S. ECA rejects a project on environmental grounds, there is still a high probability another ECA will finance it. This is what happened with financing for the Three Gorges Dam in China. Construction began on the twenty-year project in 1993. The Canadian, German, Swedish, and Swiss ECAs have all helped to finance this project, which will displace up to 2 million people and is projected to cause widespread ecological disruption, even though the World Bank and the U.S. Ex-Im Bank rejected it.[93]

The United States and the United Kingdom have pressured other G-8 countries since the late 1990s to work toward a set of environmental and social guidelines for ECAs. The OECD has followed this lead and also launched talks on common standards for ECAs in these areas. Thus far, neither forum has agreed on what these guidelines should look like.[94] At the same time, environmental NGOs have kept up their campaigns on ECAs, and in 2000 launched the Jakarta Declaration for Reform of Official Export Credit and Investment Insurance Agencies.[95] This declaration calls for greater transparency of ECAs, binding social and environmental guidelines and standards that are at least equal to those of the World Bank and the OECD, and the adoption of criteria on human rights and corruption. It calls, too, for an end to nonproductive investments and the cancellation of ECA debt of the poorest countries.

Private Finance and the Environment

The environmental impact of private global financing on developing countries has received far less attention than concessional assistance. This is understandable. Public financing is easier to track, and in the case of agencies like the World Bank, there are official policies to hold it accountable. It is much more difficult to hold private financial agencies accountable, especially when environmental standards and guidelines do not exist, as in the case of ECAs. The volume of private finance is enormous. Every day, as we noted earlier, over US$1 trillion changes hands in private financial markets, mainly currency speculation and investments in stocks and bonds. Because of its sheer size, private financing will inevitably alter the global environment, although the effects may be less obvious than in the case of public global financing.

Currency speculation and trading in particular can contribute to great economic instabilities. The 1997–1999 Asian financial crisis, for example, saw the value of local currencies collapse in countries like Indonesia. The value of the Indonesian rupiah fell from about Rp2,450 to the U.S. dollar in July 1997 to Rp11,000 to the U.S. dollar in September 1998. Unemployment and prices for essential goods soared as the Indonesian economy contracted by 10–15 percent in 1998. Economist Paul Krugman wrote in September 1998: "Never in the course of economic events—not even in the early years of the Depression—has so large a part of the world economy experienced so devastating a fall from grace."[96] Sudden and unexpected crises like the one in Asia only reinforce the short-term mentality among investors in currency markets. Similarly, the money invested in stocks and bonds through mutual funds demands short-term gains as well, so most investment ends up with firms that can promise such gains.

There is an increasing recognition both inside and outside of the financial industry, however, that this short-term mentality tends to go against the long-term vision needed by firms to promote environmental protection and sustainability. Critics worry that it increasingly makes more financial sense, for example, to harvest an old-growth forest and invest the proceeds today than it does to harvest the forest sustainably over a number of years. The returns on the former are significantly higher than on the latter. Such realities prompt firms and the banks that back them to pursue projects that lead to environmental destruction in the short run, with little consideration for the long run. By operating in this way, financial markets discriminate against firms that promote sustainable practices.[97] Unless banks and individual investors demand environmental responsibility as a condition of receiving money, there is little prospect that it will be a serious consideration. Some, such as Swiss industrialist Stephan Schmidheiny, are working to convince actors within private financial markets that they must promote principles of sustainable development.[98] He argues, along the lines of market liberals, that banks and private actors are the key to the promotion of sustainable development. If they do not act now to promote environmental protection within the private financial system, he maintains, economic-growth prospects will be harmed in the future.[99]

Over the past decade agencies like the UNEP have pushed for greater environmental accountability of banks and financial markets. The UNEP Finance Initiatives was founded in 1992 with the aim of engaging financial institutions in the broad discussions on sustainable development.[100] Financial institutions are participating in part because of NGO pressure, but in part too out of fear that financial-market investors and banks could be held liable for the environmental damage done by corporate borrowers. A 1990 legal case in the United States in fact had decided that a bank was in part responsible for the environmental damage done by one of its borrowers.[101] Such fears have been a key factor in getting banks to even consider the need to evaluate the potential environmental implications of borrowers. UNEP organized the Statement by Financial Institutions on the Environment and Sustainable Development in the mid-1990s, signed by over 260 financial institutions.[102] This statement is market friendly, yet still supports a precautionary approach to environmental management. It is not a binding agreement, but rather requests signatories to abide by its principles. This has led some to ask whether it is a genuinely green move, or just a strategy to minimize risk on the part of banks. For others, it shows the willingness of private financial institutions to take some responsibility for environmental management, knowing a healthy global environment means a healthy global economy.

Conclusion

Market liberals, then, see global financing as an effective means to promote growth and prosperity in developing countries. Grants, such as for education, food relief, and health services, can no doubt help countries adjust, reorient, and get back on the development track. There are also genuine humanitarian concerns that require grant aid. But market liberals also argue that the global community must be careful to avoid creating a welfare world, a world of unaccountable and corrupt elites living off the handouts of rich countries. It is equally critical to recognize the value of concessional loans from organizations like the World Bank, the IMF, and the Japanese government. These loans help to stabilize developing economies, which in turn benefits the entire global economy. Large loans allow governments in developing countries to lay the foundational infrastructure (especially for transportation and energy) to compete in global markets. They foster self-reliance and independence, making the long-term goal of graduating from a dependence on aid more realistic. They force recipients to take responsibility for long-term outcomes of decisions. And they demonstrate a confidence in the abilities of recipients to manage their own affairs, thus generating trust and competence within recipient countries. Multilateral and bilateral lenders can help guide the elites of developing states with expert economic advice: in particular, to explain the pitfalls of protectionism and bad financial practices. State planners in developing states need incentives to ensure that aid in fact "performs"—that is, to ensure it creates lasting economic development. Key measures of performance are the ability to service and repay loans. The financial accountability of servicing and repaying foreign debt can in addition help prod corrupt and incompetent leaders from power, which in the long run will only improve economic, social, and environmental conditions.

No doubt, market liberals concede, economic development often requires larger volumes of natural resource exports to earn the necessary foreign exchange to finance loans. Yet the period of resource exports should be seen as the first stage of development, a necessary step toward processing and manufacturing, and eventually services and information. This ratcheting-up process will inevitably involve some irreversible environmental changes. But this is not a reason to condemn natural resource exploitation. It is paternalistic to want to "deep freeze" the developing world in a past of hunters and gatherers, pristine (yet impassable) forests and mountains, and isolated (yet primitive) villages. Life for people in such conditions is short and difficult. Is it right to trap poor people in the past so rich people can save a few apes and tigers? Why should the peoples of the developing world be denied the lifestyle of peoples in the developed world? Europe and North America drew heavily on their natural resources to develop. It is nothing more than neocolonialism to prevent Asia, Africa, and Latin America from doing the same. Some of the most ardent supporters of this view are leaders in the developing world. Former Malaysian Prime Minister Mahathir has led the charge. Typical is his blunt remark in the mid-1990s: "Malaysia wants to develop, and I say to the so-called environmentalists, 'Mind your own business.'"[103]

Institutionalists agree with the broad parameters of the market-liberal interpretation of the environmental consequences of global financing. There are, however, some notable caveats. There is, institutionalists argue, a great need for *more* assistance, both loans and grants. Grants might be preferable in many situations, but to raise sufficient funds for the institutional needs of developing states and weak global environmental institutions will most certainly require concessional and nonconcessional loans as well. At a bare minimum, governments need to meet the international promise of 0.7 percent of their gross national product. There is a need, too, for donors to target more aid at specific environmental problems, especially global problems that quite naturally are a low priority in countries struggling to feed and house their own citizens. There is also a need to ease some of the debt burden, especially in heavily indebted countries, because these financial pressures weaken the capacity of these states to manage environmental resources. The focus of global financing must remain realistic, with the ultimate goal of promoting sustainable development. Economic growth and higher incomes that flow from effective global financing can then enhance the global and national capacities to manage environmental change.

Institutionalists call for more mechanisms like GEF—with innovative mandates to finance projects with global environmental benefits—because these are the only effective and practical way forward. In addition, the world needs stronger policies and loan/grant conditions to counteract some of the pressures of foreign debts (that encourage states to overexploit natural resources or overproduce cash crops). But it is just as critical to remain realistic. The developing world needs loans to survive. Without loans, developing economies would stagnate, social conditions and state services would deteriorate, and entire societies would collapse into chaos and war. There is a fine balance between the financial needs of development and the financial needs to protect the global environment. Notably, the World Bank's recent effort to mainstream the environment into lending is shifting it away from a market-liberal approach and toward an institutionalist one.

Compared to market liberals or institutionalists, bioenvironmentalists have far less to say about global financing. Most bioenvironmentalists see development lending as fuel for out-of-control growth and consumption. Environmental grants certainly have the potential to do some good, but these are a mere ripple in the global political economy, and the amounts are unable to counteract the effects of economic growth, consumption, trade, and investment. Foreign debt is, moreover, a drag on the ability of developing states to shift economic course, pushing them to harvest the last of their natural heritage just to service and repay loans.

A few extreme bioenvironmentalists—notably Hardin—call for tough measures to ensure the survival of the globe. He uses a metaphor of a world of "sovereign lifeboats." There are two kinds of boats: well-built, stable ones full of healthy survivors (the rich countries); and leaking, capsizing ones, with desperate people falling or jumping overboard, who then either beg for assistance or claw at the gunnels of the stable boats. For humanity as a whole to survive, the people in the rich boats, even the good-hearted ones, must, indeed it is their ethical duty, to let the unlucky fend for themselves—otherwise, all will drown, because there is just not enough space left (carrying capacity) in the rich boats. It is thus essential for the rich to end humanitarian assistance, since this merely prolongs the agony of those in the overpopulated waters of the poor world, and risks the lives of those in the rich world. Any other choice, in Hardin's view, is "suicidal" for "posterity." He writes: "To be generous with one's own possessions is one thing: to be generous with posterity's is quite another. This, I think, is the point that must be gotten across to those who would, from a commendable love of distributive justice, institute a ruinous system on the commons."[104] Hardin, no doubt, is at the margins of bioenvironmentalist thought, and most would see aid as a valuable tool to assist developing countries with implementing tough population-control policies. Most, too, would in the meantime throw a life preserver to at least some of the drowning (enough at least to keep them afloat). It is the overconsumption and greed of those in the stable boats that will ultimately sink the world anyway.

Social greens are the main critics of the market-liberal and institutionalist views of global financing. They also charge extremists like Hardin with moral indifference. All humans should be treated with dignity, and the accident of birthplace should not guide the morality of access and allocation of resources. The rich, social greens argue, need to care about the poor more, not less. This does not mean distributing occasional "handouts" to the poor (foreign aid). Instead, ideally, the globe needs a more just political economy, in which the rich make do with less and do not deprive the poor of sustainable livelihoods. Then there would be no need for international aid, because the poorest would have access to land and would no longer be poor. The global political economy needs to empower communities, so locals again control decisions and destinies. This will never occur within current global financing structures. The GEF, according to social greens, is an elitist institution designed to serve the interests of the World Bank and donor states. The ECAs are opaque and unaccountable, in many ways even worse than the World Bank, which is at least inching ever so slowly in the right direction. Some grant aid supports communities through education, participation, women in development, freshwater projects, sustainable employment, and so on. But the total impact of ODA is terrible. Public and private loans primarily serve global capitalism and local cronyism. The common people rarely benefit from this form of financing. The debt burden is adding to the social and environmental woes in the most destitute countries of the world. There is a critical need to waive foreign debts, and most social greens support debt-cancellation initiatives (i.e., outright forgiveness of debts of the world's most heavily indebted countries) such as the Jubilee 2000 campaign promoted.[105] Many are, at the same time, critical of World Bank efforts to ease debt burdens though programs such as the Heavily Indebted Poor Countries (HIPC) initiative, which makes limited debt relief dependent on sustained commitment to structural adjustment measures.[106] In their view, the conditions for the relief yet again mainly serve donors and their corporate and political allies rather than the peoples in developing countries.

All perspectives agree that global financing *alone* will not be enough to solve global environmental problems. What, then, is needed? The conclusion examines the heated and radically different answers to this deceptively simple question.

8

Paths to a Green World? Four Visions for a Healthy Global Environment

How, then, can the global community ensure a healthy and thriving ecological future? Should governments and global organizations control the forces of globalization? If so, what are the most effective means? Should the global community rely on technology, regimes, a world government, or local communities? There are radically different visions of the best way forward, ones rooted in radically different explanations of the causes and consequences of global environmental change. Drawing on their own unique insights, each perspective offers its own mix of proposals.

Market liberals call for reforms to facilitate a smooth functioning of markets. They want eco-efficiency, voluntary corporate responsibility, and more technological cooperation. Institutionalists call for reforms to facilitate global cooperation and stronger institutions. They call for new and better environmental regimes, changes to global organizations, and efforts to enhance state capacity to manage environmental change. Bioenvironmentalists call for reforms to protect nature from humanity. They call for lower rates of population growth and consumption as well as a new economy based on an ethic of sustainability, one that operates at a steady state, designed to preserve the globe's natural heritage. For some, this can be done cooperatively with new and far stronger institutions—a vision not much different from institutionalists. For others, conquering the human instinct to consume ecological space will require a coercive and supreme authority—perhaps a world government. Social greens call for reforms to reduce inequality and foster environmental justice. Humanity must rise up and dismantle global economic institutions to reverse globalization. The new global political economy must empower communities and localize trade and production. The new economy should respect the rights of women, indigenous communities, and the poor.

This chapter does not strive to convince you of one set of options. It highlights the key insights from each of the worldviews, and explains the proposed paths for change emanating from each. The chapter also examines the extent to which each worldview accepts or rejects the proposals of others. We will leave it to you to decide which elements of the worldviews you find most convincing and most useful.

Market-Liberal Vision

Market liberals see a future of ever-greater prosperity. Even the most superficial glance through history, they argue, confirms the great strides of humanity over the last few centuries. The past was not the idyllic paradise of goodwill and well-being that social greens so often seem to assume. It was nasty and short, full of suffering, disease, tyranny, and paranoia of "outsiders." Globalization is breaking down artificial barriers and enhancing cooperation and sympathy, which in turn promotes tolerance, democracy, and prosperity. True environmentalists, market liberals argue, should embrace globalization in all its forms: cultural, political, and most importantly economic. They should strive for a world with less government and more consumption, investment, trade, and development assistance. The immediate aim is to raise global and national per capita incomes. But the *real* aim is to ensure enough production so *all* of humanity prospers: only then will societies have the political will and the funds to implement sustainable development effectively.

Market liberals believe sustainable development is in the natural interest of business. According to the World Business Council for Sustainable Development, "Pursuing a mission of sustainable development can make our firms more competitive, more resilient to shocks, nimbler in a fast-changing world, more unified in purpose, more likely to attract and hold customers and the best employees, and more at ease with regulators, banks, insurers, and financial markets." The goal, in other words, is "sustainable human progress."[1]

Without economic prosperity, market liberals contend, it is unrealistic, indeed it is unfair, to ask nations or individuals to sacrifice in order to save the global environment. Ordinary people need jobs, wages, training, and pensions *before* they will worry about climate change and ozone depletion. Ordinary people will also, quite naturally, deplete and degrade the surrounding resources to feed, house, and sustain their families. Prosperity and better environmental quality, for market liberals, go hand in hand. A prosperous world, moreover, is a more ethical world: where people have the freedom to make dreams come true, and the freedom to lead productive and useful lives.

Specific reforms need to unleash the power of the invisible hand of the market to drive economic growth, foster innovation and efficiency, and correct distortions. This requires open and competitive markets that foster trade based on comparative advantage. It requires steady and predictable investment climates: that means stable intellectual and physical property rights, reliable contracts enforced by the rule of law, transparent and accountable standards (the elimination of state corruption), and tax reforms to avoid unfair penalties on income earners (reasonable taxes on waste and pollution are fine). It requires fewer command-and-control regulations and better market incentives to ensure "voluntary" compliance (if compliance is voluntary, but with proper incentives, it will be in the financial interest of firms to comply). It further requires global rules to prevent governments from implementing policies—like bans, subsidies, and tariff barriers—that distort the smooth functioning of markets. *The Economist* magazine explains the consequences of such policies: "Bans and rules . . . have a habit of attracting a more pernicious sort of price signal, corruption. A quicker way for a government to improve its environment is to examine its list of subsidies. Artificially cheap water and energy, as well as tax breaks for mining, are scourges of the environment worldwide."[2]

The World Trade Organization and the International Monetary Fund, for market liberals, are good models of how global cooperation can enhance economic well-being by promoting the benefits of open markets and discouraging government intervention in the economy. Further efforts are still needed, however. As the World Business Council for Sustainable Development notes, governments that impede business from doing business "and that try to take the place of business in meeting people's needs keep their people poor." The Council then adds, "Denying poor people and countries access to markets is planet-destroying as well as people destroying." As evidence, the Council points to the correlation between the Index of Economic Freedom and the Human Development Index: "Roughly, the more economic freedom, the higher the levels of human development."[3]

Market liberals like Bjørn Lomborg and Julian Simon warn, too, against premature actions to correct what might well be fictitious environmental problems. Activists and academics, they argue, are just as prone to personal and professional biases as government and corporate leaders. Groups like Greenpeace, too, often exaggerate the extent of environmental problems, on occasion because of incompetence, but primarily because this is what they do: scare the masses, fundraise, scare the masses some more. Lomborg writes, "The often heard environmental exaggeration has serious consequences. It makes us scared and it makes us more likely to spend our resources and attention solving phantom problems while ignoring real and pressing (possibly non-environmental) issues."[4] Global leaders must be careful to avoid dancing to the tune of misinformed civil societies. They must show courage and leadership, only acting when there is real scientific evidence of a problem. Market liberals do recognize that on occasion some environmental problems are severe and immediate actions are necessary. It is best, however, to minimize the problems in the first place, relying on measures like fostering eco-efficiency, voluntary corporate greening, and technological innovation. We now explore each of these in more detail.

The *eco-* in *eco-efficiency* stands for both "economic" and "ecological." By efficiency, market liberals mean we have to make optimal use of both economy and ecology. The basic idea is to achieve more from less energy and raw-material inputs. As Livio DeSimone and Frank Popoff of the World Business Council for Sustainable Development explain, "Eco-efficiency harnesses the business concept of creating value and links it with environmental concerns. The goal is to create value for society, and for the company, by doing more with less over the entire life cycle, that is, from creation of raw materials to disposal of products at the end of their life." It is "a *management philosophy* that links environmental excellence to business excellence."[5] The bottom line here is simple: benefits will accrue to companies that follow the principles of eco-efficiency. This approach will enhance their image, save them money, and gain them markets. It will also do away with the need for command-and-control regulations. It will, most importantly, ensure future growth, which is essential for the health of societies, the health of business, and ultimately the health of the planet.

Market liberals also stress the need for voluntary corporate social and environmental responsibility to ensure sustainable development. Voluntary corporate initiatives include ISO 14000 environmental management standards, the UN's Global Compact, the public-private partnerships promoted at the 2002 Johannesburg Summit, corporate social responsibility, and corporate environmental stewardship. For market liberals, adherence to these standards promises only positive outcomes for the environment. And industry is seen to be the best actor to set up such standards because it is most familiar with corporate possibilities and constraints. Market liberals prefer such an approach because it shifts the burden of regulation from the state to the firm, which can monitor environmental performance much more efficiently (important for eco-efficiency). Moreover, adherence to a global standard helps ensure that, even when state enforcement is weak (common throughout the developing world), firms will still abide by local regulations. Global standards like the ISO 14000 in addition encourage firms to adopt cleaner technologies, rather than focusing on cleaning up after the fact, because firms consider environmental impacts throughout the decision-making process. Such standards are also intended to spread good environmental practices up and down the supply chain, as well as across borders, because certified firms are required to encourage suppliers to become certified.[6]

Corporate social responsibility (CSR) is emerging as a core response of global business to the need to promote sustainable development. According to the World Business Council for Sustainable Development, it has "always stressed that sustainable development rests on three fundamental pillars: economic growth, ecological balance, and social progress. As an engine for social progress, CSR helps companies live up to their responsibilities as global citizens and local neighbors in a fast-changing world."[7] Corporate environmental stewardship is similar to CSR, but focuses more specifically on environmental conservation. It requires firms to not only voluntarily comply with all environmental laws, but to go "beyond compliance" to prevent pollution and conserve resources.[8] Though strictly voluntary, compliance with CSR and corporate environmental stewardship is assumed to be in the economic interest of business. These initiatives are thus seen to be "win-win" propositions.

Initiatives like eco-efficiency and voluntary corporate responsibility alone will not suffice to solve all global environmental problems. A sustainable future, market liberals stress, will inevitably require scientific progress, imagination, and forward thinking—the same forces that led to so many of history's great discoveries and advances. The world's wealthiest nations must nurture and harness the greatest minds—the Isaac Newtons, Henry Fords, and Albert Einsteins—for the benefit of the entire world. Societies and governments, market liberals argue, must draw on these geniuses to think us out of puzzles like climate change and deforestation. There is no other way. How, for example, given the desires and needs of ordinary people in all corners of the world to drive automobiles, can the global community possibly "solve" urban air pollution, climate change, and the inevitable end of oil with policies alone? The only feasible path is new technologies. The World Business Council for Sustainable Development is blunt here: "Recent history suggests that those living in wealthier countries do not intend to consume and waste less. Given that the other 80% of the planet's people seek to emulate those consumption habits, the only hope for sustainability is to change *forms* of consumption. To do so, we must innovate."[9]

There are many grand visions and much ongoing research on future environmental technologies, like hydrogen fuel cells for automobiles.[10] For market liberals, solutions to problems like climate change that discount the potential of new discoveries and technologies—such as calls by radicals to limit or ban cars—are naive, wishful thinking, or downright harmful to the ultimate goal of a better world. Market liberals argue that innovation that brings cleaner production processes feeds into the goal of eco-efficiency, and is thus by its nature in the interest of firms.

It is equally vital, market liberals contend, to transfer these technological advances to the developing world. The transfer of environmentally sound technology can help developing countries "leapfrog" past development stages that tend to rely on inefficient and polluting technologies.[11] Transfer of innovative environmental technologies, according to market liberals, works best when done by transnational corporations, in a business-to-business setting. Governments and international organizations should, therefore, actively entice foreign direct investment that brings in more advanced technologies.[12]

Institutionalist Vision

Institutionalists agree with many of the core recommendations of market liberals, including the need to embrace globalization, raise per capita national incomes, and facilitate trade, investment, and financing. They agree, too, that free markets, technology transfers, and voluntary corporate greening are all potentially valuable strategies to promote better global environmental management. Yet institutionalists add an important caveat to these strategies. They argue that more concerted efforts, beyond relying on markets alone, to harness and channel globalization toward sustainable development *must* accompany these strategies. Doing this will require stronger international organizations, norms, and regimes as well as stronger national and local capacities to manage environmental affairs.

Without such measures, problems like climate change *could* slow or even stall the steady progress toward a more civilized and prosperous world. The global community should not risk such a future. Like market liberals, institutionalists stress that modern scientific inquiry should guide all analyses, decisions, and agreements. At the same time it is important to recognize that the scientific method is often a slow process, involving claims and counterclaims. There is thus often great uncertainty surrounding scientific inquiry, especially for problems as multifaceted as climate change. Institutionalists see the precautionary principle as a useful tool to help overcome the limitations of scientific investigation. Principle 15 of the 1992 Rio Declaration defines this: "In order to protect the environment, the precautionary approach shall be widely applied by States according to their capabilities. Where there are threats of serious or irreversible damage, lack of full scientific certainty shall not be used as a reason for postponing cost-effective measures to prevent environmental degradation."[13] The precautionary approach has been particularly important as a justification for creating and strengthening global environmental regimes.

Environmental regimes are at the core of the institutionalist vision of a sustainable future.[14] This emphasis arises from the belief that negotiated global agreements, principles, and norms offer the best way to address environmental problems in a world of sovereign states. Global "rules" embodied in regimes provide a common point around which states can coordinate their own environmental behavior while at the same time being assured that behavior change for the global common good does not compromise their national sovereignty. Regimes also allow states to establish common but differentiated responsibilities for developed- and developing-country governments, such that states with more capacity can make greater commitments, while those with fewer resources receive less ambitious targets and more time or financial assistance to meet their commitments. This is one way for the global community to more fairly divide up responsibilities in a highly unequal world. There are several hundred international legal instruments dealing with environmental matters. This encompasses a wide range of regimes, including, for example, the 1989 Basel Convention on the Transboundary Movement of Hazardous Waste and Its Disposal, the 1992 Convention on Biological Diversity, and the 1994 UN Convention to Combat Desertification in Those Countries Experiencing Serious Drought and/or Desertification, Particularly in Africa (see table 3.1).

While institutionalists believe strongly in the potential of international environmental cooperation through regimes, many call for efforts to make them more effective.[15] Some of the more recent international environmental agreements, like the 1997 Kyoto Protocol, still need additional signatories to ratify them before they will come into force. There is also a need to ensure better monitoring and greater compliance with agreements that are already in force. And there is a need for better coordination between agreements that are related, such as between the Kyoto Protocol (climate change) and the Montreal Protocol (ozone depletion). Similarly, there could be better coordination between the Basel Convention (hazardous waste trade), the Stockholm Convention (persistent organic pollutants), and the Rotterdam Convention (hazardous chemicals and pesticides). Institutionalists also point out that there is a need for more financing, especially for the implementation of agreements in developing countries, although the Global Environment Facility, with commitments of more than US$4 billion in over 140 countries, is certainly helping. There is also a need to ensure that global agreements feed into national policies and actions.

One of the greatest concerns of institutionalists is whether regimes are in fact effective. In other words, are they actually contributing to "solving" the problem? It is extremely difficult to measure regime effectiveness, because multiple intervening factors can influence environmental change in unexpected ways. Over the past decade institutionalist scholars have shifted from their initial focus on regime formation to issues of implementation, compliance, and enforcement, which more closely reflect a regime's effectiveness. The purpose is to analyze what works and what does not in a bid to design better agreements in the future.[16]

Institutionalists also call for broad reforms to global environmental institutions. The UN Environment Programme, for example, argues:

> Many environmental institutions were originally set up under different conditions and to perform different functions from those they are expected to exercise today. . . . Many institutions are constrained by a lack of human capacity and funding, despite increased environmental challenges, and this limits their effectiveness. These are clearly issues that need to be addressed if institutions are to fulfil their present obligations and confront emerging environmental issues.[17]

Institutionalists call, too, for specific reforms to international economic organizations, as previous chapters discuss. Many people, for example, would like to see the World Bank improve its accountability by ensuring more participation of affected peoples and NGOs in both project and adjustment lending.[18] Some call for reforms to the World Trade Organization, like clarification of the relationship between multilateral environmental agreements and trade rules, as well as incorporation of the precautionary principle into the rules of the World Trade Organization (WTO).[19]

Some institutionalists, such as political scientist Frank Biermann, go further and call for the creation of a World Environment Organization (WEO) to counterbalance the economic power of organizations like the WTO.[20] The idea here is to effectively replace the relatively weak UN Environment Programme (UNEP) with a much stronger body. This could mean upgrading the UNEP from a program to a specialized agency, or it could mean streamlining a variety of activities into a larger agency to better coordinate global environmental policies. Some would even like to see an organization with enforcement powers to ensure that states actually implement and comply with the environmental agreements they sign and ratify, something UNEP currently does not have.[21] Others argue that such an organization will not necessarily improve global environmental governance and may harm the interests of developing countries. Instead they would like to see the current global environmental treaty structures enhanced.[22]

A further broad set of institutionalist recommendations focuses on the need to enhance state capacity to manage environmental issues more efficiently and effectively. This call for capacity building is geared particularly but not exclusively toward developing countries. Improved capacity at the level of the state as well as other key environmental actors can help a state to better comply with its international environmental commitments, and to improve its domestic environmental performance.[23] Meeting this goal, institutionalists argue, will require more aid for developing countries—at a minimum, developed countries should meet their promise to provide development assistance worth 0.7 percent of gross national product (see chapter 7 for details).

Capacity building is a core mandate of the UNEP and UN system as a whole. "Over the past several years," the UNEP notes, "it has become clear that capacity building is central to the quest for sustainable development."[24] Institutional and state capacity building also figured prominently in the 2002 Johannesburg Plan of Implementation, as well as in the World Bank's 2003 *World Development Report* on the theme of Sustainable Development.[25] This focus on strengthening institutional capacity marked an important addition to the World Bank's analysis of environmental issues, incorporating a much more institutionalist approach than it had in the past. Building capacity requires more than just more development assistance in terms of grants and loans. It requires technology transfers as well. Today, computers and satellites no doubt provide the most effective means of monitoring the environmental quality of resources like forests, rivers, and air.

Bioenvironmentalist Vision

Bioenvironmentalists tend to see a future of environmental doom. For many, humans are like any other animal: they act and react to survive— what some label as self-interest and others label as instinct. The human species, then, is seen to be a primary problem for the earth's ecology. The genetics that allowed the species to conquer all others in an era of seemingly limitless land and resources now leaves the species overflowing into every possible ecological space.[26] Under these circumstances, bioenvironmentalists demand radical changes to confront today's ecological crisis: the earth is beyond its carrying capacity and it is the duty of intelligent leaders to act immediately. They call for a new global political economy that respects the biological limits of the planet earth. This requires limits on the growth of the global population and the global economy. It also requires new attitudes, norms, and policies that internalize the value of ecospheres, and in particular the value of nonhuman life.

A common theme for bioenvironmentalists is the need to stem global population growth. Early bioenvironmentalists, most notably Stanford University professor Paul Ehrlich, called for stern measures to stop the "population bomb" from wiping out the planet.[27] Strong governmental powers should be granted, he argued, to coerce people into having fewer children. He advised donors to withhold food aid to developing countries unless they took drastic measures, including forced sterilization, to bring their populations under control. "Coercion?" Ehrlich ponders, "Perhaps, but coercion in a good cause. . . . We must be relentless in pushing for population control around the world."[28] Garrett Hardin, writing around the same time, called for a supreme authority to control global population growth, to implement "mutual coercion, mutually agreed upon."[29] Herman Daly, professor of ecological economics at the University of Maryland, continues this theme, advocating the issuing of transferable "birth licenses."[30] These strategies, however, now sit on the more extreme end of bioenvironmentalist thought.

Despite demographic trends in the past 35 years that suggest a slowing down of the overall rate of world population growth, more contemporary bioenvironmentalists continue to stress the need to control population growth rates.[31] The total number of humans is still on the rise, these bioenvironmentalists argue, and the planet is already beyond its carrying capacity. Today the strategies of bioenvironmentalists have softened in the face of widespread criticism over the years that employing coercive measures to control population growth violates human rights. Most bioenvironmentalists now stress the need for education, health care, and distribution of contraceptive devices that give men and women a choice of various family-planning options.[32] It is stressed, for example, that provision of universal primary education for boys and girls as well as enhanced educational opportunities at higher levels for women could go a long way toward the goal of population reduction. The idea is that educated women are more likely to pursue careers that use that education, which delays marriage and reduces family size. Enhancing health care to reduce child mortality is another humane way to encourage parents to have fewer children.

Most bioenvironmentalists tie their call for population control tightly to their call for a reduction in economic growth and consumption. These measures, say bioenvironmentalists, work equally together to deplete the earth of its natural resources. The calls to reduce economic growth and consumption have focused on bringing about fundamental changes to global consumption patterns, especially in rich countries. For this reason, some extreme bioenvironmentalists advocate controls on immigration from developing countries to rich countries. Many moderates, however, propose education as the best (effective and noncoercive) way to confront the emerging global culture of consumption. This involves questioning the implicit link between well-being and buying more and newer things. And it involves deconstructing the branding of desire to free good people from the brainwashing of advertisers.[33] The educational goal is to teach citizens to *choose* to live within limits—to use less, waste less, recycle, and alter lifestyles.[34] This approach tends to stress the role of individuals in reducing the global impact of consumption and production.

An important strategy to achieve these population and consumption goals is to promote a new economy based on a new ethic of sustainability. Daly refers to this as a "steady-state" economy, in which the number of humans and the amount of capital are constant, and at levels sufficient for a good and sustainable life. In addition, a steady-state economy has a rate of throughput of matter-energy as low as is feasible.[35] Under such an economy, there is room for society to develop, to improve human well-being, but not to grow. Daly is firm, though, in stating that a steady-state economy does not imply zero growth in gross national product, because it is not defined in terms of gross national product (GNP).[36]

Some bioenvironmentalists argue that one way we can instill this new sustainability ethic and move toward a steady-state economy is to do away with our measures of "progress" and "well-being," in particular the misleading focus on gross domestic product and gross national product. There are a number of alternative economic indicators that bioenvironmentalists propose, some of which focus on adjusting GDP and GNP figures, and others that focus mainly on environmental indicators. Two of the more prominent indicators are the index of sustainable economic welfare (ISEW) and the genuine progress indicator (GPI). The ISEW, first proposed by Herman Daly and John Cobb,[37] measures real per capita personal consumption expenditure, and adjusts it to correct for income inequality, pollution, loss of natural capital, value of household labor, and a variety of other factors not counted in GNP or GDP. The GPI, as proposed by the U.S. NGO Redefining Progress, measures the financial transactions from GDP that relate to well-being and adjusts them for factors similar to those incorporated into the ISEW.[38] Both indicators are trying to measure roughly the same thing, human well-being beyond mere GDP, but they have slightly different methodologies. A comparison of the ISEW and the GPI against GDP growth over the past 50 years shows clearly that while GDP has continued to grow, the other indices show very little improvement.[39] The ISEW and GPI have gained important recognition among environmental groups such as Friends of the Earth, and they have now been calculated for a number of countries.[40]

Indicators promoted by bioenvironmentalists that focus mainly on environmental measures include the ecological footprint (see chapter 4) as well the living planet index (LPI). The ecological footprint has been used by organizations such as the WWF to demonstrate the extent of planetary resource depletion. The WWF also developed the LPI, which averages changes on three indices: forests, freshwater ecosystems, and marine ecosystems.[41] Such measures are not designed to replace GDP or GNP, but rather to supplement them. The aim is to demonstrate not just to individuals but also to governments that we are living beyond the planet's biological capacity. The WWF's analysis of ecological footprints, for example, shows that human consumption of natural resources in 1999 "overshot the Earth's biological capacity by about 20 per cent."[42] The aim is to use this information to change human behavior in ways that are less consumptive and more conducive to conservation.

Bioenvironmentalists also commonly call for global restraints on production, trade, and especially transnational corporations as a way to move toward a more sustainable global economy. Some call for a world government or stronger international authority to ensure that such restraints are put in place. Others instead call for a constellation of strong institutions, norms, and rules—from global to local and public to private—to constrain production, trade, and transnational corporations so humanity again lives within the carrying capacity of the earth. Many recognize, however, that this will be a difficult challenge given what is seen to be humanity's innate drive to plunder the planet.[43] Measures along these lines called for by bioenvironmentalists are similar to the institutionalist proposals discussed earlier, but bioenvironmentalists add (and stress) that a fundamental shift in human values must first occur.

Overall, then, bioenvironmentalists see a need to profoundly change Western values and lifestyles to solve global environmental problems. We must move, as Thomas Princen of the University of Michigan's School of Natural Resources and the Environment argues, from values based on efficiency and cooperation alone to ones focused on "sufficiency." In other words, accept "enough" and stop aiming for "more" through the growth paradigm.[44] There is no escaping this—not even with Nobel Prize–winning technological breakthroughs. How, these bioenvironmentalists ask, can technology ever replace an unidentified species? We cannot replace what we never knew we lost.

Social Green Vision

Social greens want a world that ensures both social and environmental justice. This is not just an ethical stance. Without this justice, social greens argue, the resulting global ecological crisis will swamp us all. Social greens largely agree with bioenvironmentalists on the need to address the culture of overconsumption and the need to adopt a new global ethic. Their focus, however, is much more squarely on rejecting globalization and relying on citizen action to promote justice in local communities. Political and environmental scientist Michael Maniates, for example, provocatively argues that the "individualization of responsibility" for environmental choices—such as green consumption and recycling—can feed into corporate goals and the culture of consumption. This can appease the guilt of overconsumption and diffuse the societal will to imagine collective ways to tackle environmental threats. He makes the important, and often overlooked point, that "individual consumption choices *are* environmentally important, but . . . control over these choices is constrained, shaped, and framed by institutions and political forces that can be remade only through collective citizen action, as opposed to individual consumer behavior."[45]

How can the global community bring about social and ecological justice in ways that reduce overall global consumption? Social greens call on the global community to dismantle or downgrade global economic institutions, change the nature of trade and production, cancel developing-country debts, and localize economies to empower local communities—to, in effect, reverse globalization and industrial capitalism and alter the nature of the global political order. We now address each of these proposals.

In terms of global trade rules, most social greens argue that global leaders and ordinary citizens should reject the latest round of WTO trade negotiations, at least until a full environmental assessment of WTO activities is conducted.[46] Many, such as the members of the International Forum on Globalization (IFG), would like to see the WTO rules rolled back and the institution dismantled. Activist Colin Hines suggests a Global Agreement on Sustainable Trade (GAST) to replace WTO rules, as well as a World Localization Organization (WLO).[47] He envisions that such organizations would encourage states to give preferential treatment to goods and services from other states that respect human rights, treat workers fairly, and protect the environment, and would favor domestic industry over foreign goods. Others, such as Greenpeace, take a more pragmatic approach and call for clear incorporation of the precautionary principle into WTO rules. They call, too, for precedence of multilateral environmental agreements (MEAs) over WTO rules, as well as greater WTO transparency and consultation. They also recommend rules to allow for discrimination based on production and processing methods (PPMs) and rules against patents on life forms (elements of what is sometimes called "safe trade").[48] Where trade is unavoidable, social greens also call for "fair trade"—that is, small-scale trade with communities that ensures fairer prices for goods and environmentally sound production.[49]

On global financing, social greens call on donors to forgive foreign debt as well as create a new organizational structure for global financing. Among social greens there is a split between those who want to dismantle the existing global economic infrastructure, and those who call for significant reforms. Most social greens, it is important to note, are not against global economic rules per se. The IFG, a group of more radical social greens, states: "Clearly, there is a need for international rules. To serve the whole of humanity, however, they must be based on the consent of the governed, and enforcement must be left primarily to democratically elected local and national governments."[50]

Many, both radicals and reformists, would like to see the International Monetary Fund and World Bank decommissioned and favor a rollback of all structural adjustment programs.[51] A new organization, providing financial advice but not loans, could work toward meeting goals of debt relief and local sustainability. Some would like to see international funding in the form of loans only available on a regional scale, and only for short-term emergencies.[52] Some envision an International Insolvency Court (IIC), which would oversee debt forgiveness. The New Economics Foundation and the Jubilee Coalition have proposed a Jubilee Framework, somewhat more reformist in nature, to oversee debt relief.[53] Many social greens also propose some sort of Tobin Tax—that is, a small tax of less than half a percent—on global financial transactions as a way to dampen global financial speculation, with the proceeds going to debt relief and/or local green initiatives.[54]

With respect to TNCs and global investment, most social greens advocate some sort of global agreement on corporate accountability to curtail what they see as systemic environmental and social abuses by TNCs. Most argue that voluntary measures are wholly inadequate, and that a legally binding agreement with teeth is required. Several proposals have been put forward by social green organizations. Greenpeace, for example, is promoting the "Bhopal Principles on Corporate Accountability." This includes stricter liability for corporations and rules to ensure the application of the highest standards regardless of location.[55] Although this proposal calls on firms to adopt these principles, their ultimate aim is for a legally binding treaty along these lines. Friends of the Earth International, too, has called for a legally binding treaty on corporate accountability.[56] Some social greens go so far as envisioning a UN organization to oversee such an agreement, as well as to collect data on TNC activities.[57]

The logical other side of the antiglobalization stance of social greens is the promotion of localization, also referred to as "subsidiarity." This means scaling down from the global where possible, and can refer to several different sublevels. As the IFG states, "Depending on the context, the local is defined as a subgroup within a nation-state; it may also be the nation-state itself or, occasionally, a regional grouping of nation-states. In all cases, the idea is for power to devolve to the lowest unit appropriate for a particular goal."[58]

All the institutional changes mentioned above are intended to work toward this goal of localization. Only those issues that require truly global coordination, such as global health and environmental issues, should have some level of international cooperative decision making, but others, like most economic, cultural, and political decisions, should be local. In practical terms this means promotion of small-scale local agricultural systems and local production through small-scale enterprises. Corporations would be bound by "site here to sell here" policies.[59] It also means promoting local currencies, such as local exchange trading systems (LETS).[60] Some trade may still be necessary, but social greens stress that local communities should not be overly dependent on it. All of these localization measures would preserve diversity and reinvigorate local economies and communities. "Long-term solutions to today's social and environmental problems," writes social green Helena Norberg-Hodge, "require a range of small, local initiatives that are as diverse as the cultures and environments in which they take place."[61]

A final general set of proposals from social greens call for measures to reempower voices marginalized in the process of economic globalization. Those who have been silenced include indigenous people, women, and the poor, who have been displaced from their traditional lands. These groups, social greens argue, are holders of critical ecological knowledge, and must be given the space and voice to apply it—that is, to practice sustainable livelihoods and educate others. Social greens call for local communities to reclaim the commons, including land, knowledge, water, biodiversity, and so on.[62] Reinstituting traditional and indigenous systems to protect and control common resources is, social greens contend, the most promising path to ensure a just world and a clean environment. This would involve the valuing of nonmonetized activities as legitimate economic activities, similar (but less formally measured) to the bioenvironmentalist measures of progress. Social greens see land reform as vital to enable this vision, because the world's poor and displaced peoples need fair and just access to land.[63] Social greens also lend their support to global social movements working to promote local and democratic movements to "reclaim the commons." This includes global movements for food security, land reform, indigenous rights, and sustainable local economies. Nicolas Hildyard, a former *Ecologist* editor and member of the Corner House, an NGO that works to support local communities in their environmental struggles, puts it succinctly: "Ultimately, it is only through the direct and decisive involvement of local peoples and communities in seeking solutions to the environmental crisis that the crisis will be resolved."[64]

Clashing Visions?

How compatible or incompatible are these four visions? Market liberals see potential pitfalls with the reform efforts of institutionalists, worrying their proposals to harness globalization with government and intergovernmental policies could harm economic growth. To meddle too much with the market, even if well intentioned, will result in inefficiencies and slow down progress toward better environmental conditions. In general market liberals want less government, not more, and many are skeptical of global institutions such as the United Nations. Many market liberals are also critical of global environmental treaties, arguing that in too many cases the rules of these agreements contradict the rules of the WTO, which should guide the world economic order.

At the same time, market liberals consider the antiglobalization proposals of social greens and bioenvironmentalists as wrongheaded and downright dangerous to global well-being. Shutting down economic globalization will, in the view of market liberals, cause far more environmental harm (linked to increased poverty) than growth. Moreover, they argue that the bioenvironmentalists and social greens are far too pessimistic about the benefits of new technologies, which in the view of market liberals will reverse most growth-related environmental damage. They agree with social greens that population is not as grave an issue as bioenvironmentalists claim. But their reasons are different. Market liberals see economic growth as the key mechanism by which to ease population growth. Many market liberals also share some of the social green skepticism of global institutions. Yet they see the social green proposals for localism as nothing more than a recipe for protectionism, endemic poverty, and parochialism.

Social greens and bioenvironmentalists have strong alliances in some respects, because both see the so-called solutions of market liberals as little more than greenwash. They are opposed to large-scale global markets and are highly skeptical of market-based measures at this level. They have very little faith in voluntary corporate measures. According to these thinkers, standards such as ISO 14000, for example, have done little in practice to transfer clean technologies or improve the global environment. They also think it is dangerous to rely on Western science to generate new technologies, which in the long run could aggravate rather than solve global environmental problems. But while social greens oppose large, global markets, they do not oppose the use of small-scale markets in the local context (which market liberals also advocate).

Both social greens and bioenvironmentalists see institutionalist proposals for global environmental cooperation through regimes as little more than bailing out a few buckets of water as the Titanic sinks, because global environmental agreements have done little to improve the state of the global environment. Yet social greens and bioenvironmentalists interpret the reasons for this failure differently. Bioenvironmentalists think the authority of global regimes is too weak to be effective. Social greens think that focusing on global regimes is a misguided strategy because such efforts do not adequately include the voices of those most affected by environmental change, which means these agreements are inevitably ineffective.

Bioenvironmentalists and social greens disagree sharply at times over other issues. Social greens are often critical of the bioenvironmentalist calls for coercive measures to reduce population growth or to curb immigration to rich countries. A recent split within the U.S. Sierra Club (and as well between the U.S. and Canadian arms of the organization) illustrates this division nicely.[65] A slate of candidates ran in 2004 for election to the U.S. Sierra Club Board calling for a crackdown on immigration into the United States on ecological grounds. Ben Zuckerman, a member of the Sierra Club's national board of directors, and a professor of physics and astronomy at the University of California, Los Angeles, argues that immigrants cause great national, and ultimately global, ecological harm, because allowing a poor person into a rich country like the United States dramatically increases that person's ecological footprint.[66] Opponents within the Sierra Club called such views "racist," no different from the views of white supremacists.[67] Such views led the Canadian arm of the Sierra Club, which takes a more social green perspective on population and immigration issues, to stress that it is a separate organization from its U.S. counterpart, and to develop its own population policy.[68]

Some social greens also take issue with bioenvironmentalists over the quantification of well-being in alternative measures of progress, arguing that they are merely replacing old numerical biases with new ones. Many bioenvironmentalists, too, while agreeing with many of the social green critiques of globalization, still think that global cooperation, perhaps through a higher authority, is nevertheless essential (with the important caveat of a new global sustainability ethic). Thus, many bioenvironmentalists are not keen on a localization strategy that minimizes global institutional coordination.

Institutionalists, on the other hand, tend to see little harm in the solutions of market liberals—for them these efforts will certainly help—but *only if* institutional safeguards are put in place to ensure that markets function in the best interest of the environment. They take issue, however, with those market liberals who criticize environmental treaties for contradicting the trade rules under the WTO, seeing such views as potentially dangerous for global environmental cooperation. For many institutionalists, international environmental laws should have as much standing as trade laws. Institutionalists also agree with bioenvironmentalists on the need to manage population growth more effectively. Their solutions, however, are more incremental and less coercive, focusing mainly on international cooperative efforts, through aid and technical advice, to enhance education, especially for women. Institutionalists also tend to support the general social green view on the need to rebuild the social structures of local communities. This should occur, however, through a globally coordinated effort, such as the local Agenda 21 action plans in communities tied to the global Agenda 21. Institutionalists also tend to see bioenvironmentalists as too pessimistic about the future, and see the social green vision of a world economic order of local communities as naive. They do, however, tend to see little harm in social greens' promotion of fair trade.

So what does all this mean for global environmental policy? Many would argue that what we have in practice today in both the global and national arenas are policies that largely follow the market-liberal and institutionalist worldviews. The powerful global economic institutions and some key states—most notably the United States, but also including at times Australia, New Zealand, Japan, and Canada—have adopted a more or less market-liberal worldview and are eager to see it influence global environmental governance. This is reflected in the current thrust from these players toward further economic globalization. The WTO's push for further liberalization of trade in agriculture and investment, and the continued push for programs of structural adjustment by the International Monetary Fund and the World Bank, are prime examples of the market-liberal influence. The global community is at the same time forming and extending more and more global environmental regimes, building a thick institutional structure of global environmental governance. These efforts are supported by organizations such as the UNEP, as well as by some key states (most notably those in Europe, many developing countries, and sometimes Canada). In some cases, however, such as with the Kyoto Protocol, there are formidable stumbling blocks to the successful adoption and implementation of these agreements as states with a more market-liberal stance resist them. The social greens and bioenvironmentalists offer their proposals as an alternative to this reality.

At the Johannesburg Summit in 2002 there were strong alliances between social greens and bioenvironmentalists on the one hand, and institutionalists and market liberals on the other during the debates over the ecological impacts of globalization and economic growth. The proposals of the market liberals and institutionalists, however, clearly had the upper hand in the formal proceedings, although bioenvironmentalists and social greens did manage to influence the formal agenda, such as the proposal for a global treaty on TNCs, which did generate discussion and debate. At Johannesburg, as at Rio, proposals for change adopted by the global policy community appeared to be converging around moderate institutionalist thought, because these views seem to offer a compromise between market-liberal and more radical ecological views. Institutionalists, too, seem keen to broker this compromise. The recent publications of the UNEP, for example, often canvass the arguments of market liberals, bioenvironmentalists, and social greens, drawing on different facets to promote their own agenda on the need to strengthen capacity and institutions. The trend toward institutionalist thought within the global policy community *does not* mean institutionalists have "found" the answers and solutions. All it means is that they are managing, for the moment anyway, to muster the broadest base of political support by proposing "compromise solutions" to exceedingly difficult problems. Such small sacrifices, bioenvironmentalists and social greens continue to fret, may only slow down, but not reverse, the current crisis. More extreme market liberals, on the other hand, find these efforts to be constraining (although moderate market liberals are more likely to support modest institutional compromises). It remains to be seen how extensively and effectively the global community will implement the ideas and proposals of this institutionalist compromise.

No doubt, wading through the various worldviews on the political economy of the global environment was bewildering at times. There are no straightforward trends, explanations, or *solutions*. Our choices have multiple global impacts, some good, some bad, some comprehensible, others not. There is therefore no trouble-free path to a green world. Some global environmental change is inevitable. No one disputes that. Understanding such grand change means mistakes in analysis and policies and personal choices are inevitable, too. Some days, we must admit, we wish the world was not such a dense thicket of political, socioeconomic, and ecological complexity and uncertainty. On many days we wish there were in fact simple solutions—predictable, fair, and painless. It would certainly make our own everyday choices easier. Should we eat a conventionally grown banana from Ecuador? Buy a shirt made in China? Drive our children to school? Turn on a lamp to read this book? Simple answers and simple solutions would undoubtedly allow us all to feel less hypocritical about our everyday decisions. Yet from the knowledge of the complexity of the political economy of global environmental change comes perhaps the most important lesson of this book: the need to comprehend, tolerate, and yes, even respect, the views of others as each of us develops our own vision of how to best move forward to create a healthy and prosperous planet. Embracing such diversity of knowledge, of knowledge with many "correct" answers and no "absolute" certainties, will not, we sincerely hope, paralyze environmentalists with perplexity. Instead, we hope it will empower environmentalists—from market liberals to institutionalists to bioenvironmentalists to social greens—to probe their own beliefs, so as to one day have a humility of mind that comes from a true understanding of the arguments and evidence of others.

Notes

Chapter 1

1. Other overviews of global environmental politics that complement this book include Switzer 2004; Lipschutz 2003; Maniates 2003; DeSombre 2002; Paterson 2000b; Porter, Brown, and Chasek 2000; Conca and Dabelko 1998; Elliot 1998; Dryzek and Schlosberg 1998; Dryzek 1997.

2. We use the term *political economy* to mean simply the interactions between political and economic processes. We are not implying any particular theoretical bias by using this term.

3. WCED, 1987, 43.

4. "Our Durable Planet," 1999, 29.

5. World Bank, 1992, 30.

6. Simon 1981, 1996. The late Julian Simon was an economist and taught business administration at the University of Maryland. He became well known for his work on population growth and resource use, arguing that humanity is not facing a future of global scarcities of natural resources. He published widely on these themes, including books such as *The Ultimate Resource* (1981) and *Population Matters* (1990).

7. Easterbrook 1995. Gregg Easterbrook is a senior editor of the *New Republic* and a contributing editor to the *Atlantic Monthly* and the *Washington Monthly*. In his 1995 book, *A Moment on Earth: The Coming Age of Environmental Optimism*, he argues that the global environmental situation is improving rather than deteriorating.

8. Lomborg 2001. Bjørn Lomborg is a professor of political science at the University of Århus in Denmark and director of Denmark's National Environmental Assessment Institute. Lomborg's 2001 book, *The Skeptical Environmentalist*, draws on statistics on environmental quality to argue that, overall, the state of the global environment is in far better shape than ecologists and activists claim. This book sparked considerable controversy. The Danish Committee on Scientific Dishonesty, in response to complaints that Lomborg misused statistics and exaggerated points, ruled that the book was dishonest and contrary to standards

of good scientific practice. The Danish Ministry of Science overturned this ruling in December 2003.

9. Bhagwati 2002. Jagdish Bhagwati is a professor of economics at Columbia University in New York. He has also served as Economic Policy Advisor to the Director-General, GATT, as Special Advisor to the UN on Globalization, and as External Advisor to the WTO. Bhagwati has published scores of books and articles arguing the case for free and open markets, including *In Defense of Globalization* (2004).

10. Schmidheiny 1992. Stephan Schmidheiny, a Swiss industrialist, was the founding head of the Business Council on Sustainable Development (which later became the World Business Council for Sustainable Development). His 1992 book, *Changing Course: A Global Business Perspective on Development and the Environment*, argues that it is in the economic interest of businesses to green themselves. Schmidheiny also played a central role in organizing business interests at the Rio Earth Summit in 1992.

11. Haas, Keohane, and Levy 1993.

12. Neumayer 2001, ix.

13. Paehlke 2003.

14. Young 1989, 1994, 1999, 2002. Oran Young is a political scientist at the University of California. He has published many books and articles on the theme of institutional cooperation for the environment and is known especially for his work on international environmental regimes.

15. UNEP, 2002a.

16. Keohane and Levy 1996.

17. See, for example, Victor, Raustiala, and Skolnikoff 1998.

18. Lovelock 1979, 1995. James Lovelock is an independent scientist and environmentalist. Many see his concept of Gaia as one of the most imaginative and significant contributions to understanding the global environment.

19. Rees 2002, 249. William Rees is a professor of human ecology and environmental planning at the University of British Columbia in Canada. He has published widely on issues of consumption and has developed, with Mathis Wackernagel, the concept of the "ecological footprint," which measures human impact on the earth in terms of the amount of land needed to sustain lifestyles (see chapter 4 for details).

20. Rees 2002, 249.

21. Malthus 1798.

22. Ehrlich 1968. Professor Paul Ehrlich is a biologist at Stanford University. He is best known for his work on population issues—many consider him a modern-day Thomas Malthus. He argued in his 1968 book, *The Population Bomb*, for example, that human overpopulation was destroying (or would soon destroy) the natural environment and create widespread famine and social instability. For this work he is sometimes referred to as the "prince of doom." Chapter 3 discusses his work in more detail.

23. Hardin 1968. Garrett Hardin, author of 27 books and over 350 articles, was a professor of human ecology at the University of California, Santa Barbara. His work has been widely cited, and articles like "The Tragedy of the Commons" (1968) and "Living on a Lifeboat" (1974) have appeared in hundreds of anthologies. He was a member of the Hemlock Society, believing in the right to plan his own passing, and in 2003 he committed suicide at the age of eighty-eight. He was joined by his wife, Jane, who was suffering from Lou Gehrig's disease. His many books include *Exploring New Ethics for Survival: The Voyage of the Spaceship Beagle* (1972), *Filters against Folly: How to Survive Despite Economists, Ecologists, and the Merely Eloquent* (1985), *Creative Altruism: An Ecologist Questions Motives* (1995), and *Living within Limits: Ecology, Economics, and Population Taboos* (2000).

24. Myers 1997, 34; see also Myers 1979.

25. Myers 1997.

26. Daly 1999, 34.

27. Herman Daly is a Senior Research Scholar at the University of Maryland. He served as Senior Economist at the Environment Department of the World Bank in the early 1990s, a post from which he resigned. Daly is best known for his promotion of ecological economics—economics that takes the physical limits of the planet into account. He argues that economies must attain a "steady state" that is in balance with the physical environment (for details see chapter 4). He was a founder of the International Society for Ecological Economics and has published widely on the concepts of growth and environment-trade linkages. He received the Right Livelihood Award (alternative Nobel Prize) in 1996. Among his many books are *Toward a Steady-State Economy* (1973) and *For the Common Good: Redirecting the Economy toward Community, the Environment, and a Sustainable Future* (with John Cobb, 1989).

28. Hardin 1974; Ophuls 1973, 210.

29. For an overview of the origins of these views, see Helleiner 2000 and Laferrière 2001.

30. Schumacher 1973; Sachs 1999.

31. McMurtry 1999; see also Korten 1999.

32. Paterson 1996; Levy and Newell 2002. See also Stevis and Assetto 2001.

33. Shiva 1989; Mies and Shiva 1993. Vandana Shiva is a scientist, philosopher, and feminist whose activism and scholarly writings on the themes of global agriculture, the environment, and women have won her international acclaim. She is best known for her work on intellectual property rights over seeds, and the impact of the WTO on agriculture, women, and the environment in the developing world. She has been director of the Research Foundation for Science, Technology and Ecology, based in India, since 1982. She is also a leading member of the International Forum on Globalization and is advisor to many organizations, including the Third World Network. Her many books include *Staying Alive: Women, Ecology and Development* (1989), *The Violence of the Green Revolution* (1992), *Biopiracy: The Plunder of Nature and Knowledge* (1997), and

Stolen Harvest: The Hijacking of the Global Food Supply (2000). She received the Right Livelihood Award in 1993.

34. See, for example, Princen, Maniates, and Conca 2002.

35. Sachs 1999. Wolfgang Sachs is a scholar and researcher at the Wuppertal Institute for Climate, Environment, and Energy in Germany. He is also a former chair of Greenpeace Germany. Sachs is best known for his work on the interaction of globalization, development, and the environment, in which he focuses on the role of power in the global economy and its implications for the developing world and for the environment. His books include *Global Ecology: A New Arena of Political Conflict* (1993) and *Planet Dialectics: Explorations in Environment and Development* (1999).

36. Goldsmith 1997. In 1969 Edward Goldsmith founded the magazine *The Ecologist*, which today remains a leading outlet for social green analysis. Goldsmith is coauthor of *The Social and Environmental Effects of Large Dams* (with Nicholas Hildyard, 1984), coeditor of *The Case against the Global Economy: and for a Turn toward the Local* (with Jerry Mander, 1996), and author of *The Way: An Ecological World-View* (1992). He received the Right Livelihood Award in 1991.

37. Mies and Shiva 1993, 278.

38. *The Ecologist*, 1993, 140–150.

39. See, for example, Mander and Goldsmith 1996.

40. Shiva 1993b, 150.

41. International Forum on Globalization, 2002.

42. Hines 2000, 2003. Colin Hines is an associate of the International Forum on Globalization and is coordinator of "Protect the Local, Globally," a pro-local and anti–free trade think tank. He also served as coordinator of the Economics Unit at Greenpeace International. He has published widely on the theme of promoting localization and argues against free international trade.

43. Shiva 1997, 8.

Chapter 2

1. For critical inquiry into the extent of contemporary globalization, see Scholte 2000; Hirst and Thompson 1999; Garrett 1998; Weiss 1998.

2. Scholte 1997, 14.

3. A telephone mainline is one that connects a subscriber to the telephone exchange equipment (ITU, 2003).

4. UNDP, 1999.

5. ITU, 2003.

6. UNEP, 2002b, 20.

7. UNDP, 2001, 32.

8. UNEP, 2002b, 36.

9. Held et al. 1999, 170.

10. Scholte 1997, 17.

11. UNDP, 1999, 25.

12. UNCTAD, 2002, xv, 272; UNCTAD, 2001, 9.

13. WTO, 2001.

14. Note, for example, the formal name of North Korea: Democratic People's Republic of Korea (DPRK).

15. Some of these democracies are considered weaker than others. Some organizations are devoted to tracking and ranking democracies around the world. See, for instance, World Audit at www.worldaudit.org.

16. See, for example, International Forum on Globalization, 2002.

17. UNEP, 2002b, 35, 37.

18. Held et al. 1999, 344.

19. UNDP, 2001.

20. UNDP, 2001; World Bank data profiles, available at www. eto.org.uk/eustats/graphs/93-99.htm#pcs; World Development Indicators Online: www.worldbank.org/data.

21. UNDP, 2001, 40.

22. World Tourism Organization, Tourism Market Trends 2000, at www. world-tourism.org.

23. Dauvergne 2003.

24. UNEP, 2002b, 24.

25. Helliwell (2002, 78) uses the terms *globaphiles* and *globaphobes* (from Burtless et al. 1998) to stress these contrasting views of globalization. He writes that "globaphiles are convinced that universal market openness is the single vital key to higher living standards. Globaphobes, by contrast, regard globalization as the tool multinational corporations are using to rob the world's poor by exploiting their labour, resources, and environments; destroying their culture; and commanding their vassal governments to implement whatever laws and trade agreements would make these transfers easier to achieve."

26. World Bank, World Development Indicators Online: www.worldbank. org/data.

27. UNCTAD, 2000.

28. UNEP, 2002b, 329.

29. Lomborg 2001, 4, 7.

30. Lomborg 2001, 5, 61.

31. FAO Committee on World Food Security 2001, at www.fao.org/ docrep/meeting/003/Y0147E/Y0147E00.htm#P79_3644.

32. FAO statistical database for nutrition, at www.fao.org.

33. Sen 1981.

34. Simon 1996, 12.

35. World Bank, 1992, 29.

36. Runge and Senauer 2000.

37. Paarlberg 2000.

38. Today the thickness of ozone over the Antarctic is generally 40–55 percent of its pre-1980 level (UNEP, 2000, 5). For details on the environmental politics of the Antarctic, see Elliot 1994; Joyner 1998.

39. UNEP, 2000, chap. 2.

40. Soroos 1997, 169.

41. Quoted in Wilson 2002, i.

42. Wenz 2001, 5–6, quote on 6.

43. Pojman 2000, 1.

44. Brown, Gardner, and Halweil 1999, 17.

45. Population Division of the Department of Economic and Social Affairs of the UN Secretariat, 2001.

46. UNEP, 2002b, 35. See Paterson 2000a for an analysis of the environmental politics of the global culture of cars.

47. UNFPA, 2001, 45.

48. UNFPA, 2001, 1, 6–7.

49. WMO, 1997, 9. See Gleick, Burns, and Chalecki 2002 for an overview of the technological, political, and economic forces behind the global freshwater crisis.

50. UNFPA, 2001, 5.

51. UNEP, 2001b.

52. Worldwatch Institute, 2003; quote from the summary, at www.worldwatch.org/press/news/2003/01/09.

53. Thomas et al. 2004, 145–148.

54. Pounds and Puschendorf 2004, 108. For details on the global politics of biodiversity, see Steinberg 2001; Mushita and Thompson 2002.

55. WRI, 1997.

56. Princen, Maniates, and Conca 2002, 6.

57. See, for example, Mander and Goldsmith 1996.

58. Conca 2001; Dalby 2004.

59. International Forum on Globalization, 2002, 5. The "consensus" authors with the International Forum on Globalization (IFG) are like a "Who's Who" of social greens. They are Jerry Mander, John Cavanagh, Sarah Anderson, Debi Barker, Maude Barlow, Walden Bello, Robin Broad, Tony Clarke, Edward Goldsmith, Randy Hayes, Colin Hines, Andy Kimbrell, David Korten, Sarah Larrain,

Helena Norberg-Hodge, Simon Retallack, Vandana Shiva, Vicky Tauli-Corpuz, and Lori Wallach.

60. UNEP, 2002b, 62.

61. FAO website: www.fao.org.

62. WHO, 2002.

63. Allison et al. 1999.

64. Available at www.surgeongeneral.gov/topics/obesity/calltoaction/toc.htm.

65. See UNAIDS, at www.unaids.org.

66. Karliner 1997.

67. Shiva 2000; Kneen 1999.

68. Shiva 1997; Brac de la Perrière and Seuret 2000.

69. International Forum on Globalization, 2002, 10.

70. Two books (Grundmann 2001; Parson 2002) dispute the common claim that Dupont only began to support global regulation of CFC production and emissions because it had developed a substitute. Instead, these books argue that "by 1986 Dupont, concerned about its reputation, felt the weight of the scientific evidence rendered its position untenable and, furthermore, that industry leaders feared the prospect of litigation or unilateral action by the US" (Layzer 2002, 119).

71. IPCC, 2001.

72. Meyer-Abich 1992.

73. Paterson 2001a. For more on the global politics of climate change, see Rowlands 1995, 2000; Paterson 1996, 2001b; Soroos 1997, 2001; Newell 2000; Skjærseth and Skodvin 2001.

74. Rees and Westra 2003.

Chapter 3

1. Dryzek 1997.

2. For example, Hurrell and Kingsbury 1992; Brenton 1994; Caldwell 1996; Tolba and Rummel-Bulska 1998.

3. Such problems pushed some states to cooperate—for example, Canada and the United States signed the Migratory Birds Treaty in 1918.

4. Schreurs 2002, chap. 2.

5. Crosby 1986.

6. Grove 1995.

7. Grove 1995; Fairhead and Leach 1995; Bryant 1997. Today, many ecologists recognize the logic of swidden farming in sparsely populated tropical areas. The burning of the degraded forest cover and vegetation supplies the soil with essential nutrients (acting like a natural fertilizer). After harvesting, the soil needs time

to recover, so if there is opportunity it makes sense to move on and cultivate a new area.

8. Adams 1990, 16–20.

9. Grove 1995, 18–19.

10. Adams 1990, 20–21. The IUCN shortened its name to IUCN–The World Conservation Union in 1990.

11. The IUCN's web address is www.iucn.org/.

12. Lear 1997, 416–417.

13. Jasanoff 2001, 309–337.

14. World Bank World Development Indicators Online: www.worldbank.org/data.

15. Rapley 2002.

16. See, for example, Frank 1967; Rodney 1972.

17. Helleiner 1994.

18. Ehrlich 1968, 132.

19. Hill 1969, 1–2.

20. Meadows et al. 1972. Twenty years later Donella Meadows, Dennis Meadows, and Jørgen Randers published *Beyond the Limits: Confronting Global Collapse, Envisioning a Sustainable Future*, which revisits many of the arguments in *Limits to Growth*.

21. Meadows et al. 1972.

22. See, for example, Maddox 1972; Simon 1981; Commoner 1990.

23. Meadows, in Zacher 1993, 10.

24. Kumar 1996.

25. Schumacher 1973.

26. See Rees 2002. Rees (pp. 249, 267) goes even further and argues that "a genetic predisposition for unsustainability is encoded in certain human physiological, social and behavioral traits that once conferred survival value but are now maladaptive. . . . Homo sapiens will either rise above mere animal instinct and become fully human or wink out ignominiously, a guttering candle in a violent storm of our own making." See also Brown, Gardner, and Halweil 1999; Brown 2003.

27. UN Resolution 2398, quoted in Caldwell 1996, 58.

28. Rowland 1973, 49–51. The 1971 *Founex Report on Development and Environment* is available at www.southcentre.org/publications/conundrum/annex1.pdf.

29. Caldwell 1996, 68; also see Brenton 1994.

30. Cited in D'Amato and Engel 1997, 14.

31. Cited in Brenton 1994, 39.

32. Adams 1990, 38.

33. Murphy 1983; Hoogvelt 1982.

34. Taylor 1995.

35. Akula 1995; Rangan 1996.

36. IUCN, 1980.

37. Chapter 7 discusses the environmental effects of structural adjustment in greater detail. For a good overview of the economic situation in this period, see Rapley 2002.

38. WCED, 1987, 8.

39. Bernstein 2001.

40. See Porter, Brown, and Chasek 2000.

41. See, for example, Krueger 1999; O'Neill 2000; Clapp 2001.

42. Rogers 1993, 238–239.

43. The Rio Declaration and Agenda 21 were published in UN, 1992. Also available at www.un.org/esa/sustdev/documents/agenda21/index.htm.

44. Brack, Calder, and Dolun 2001, 2. The full name of the document is the "Non-legally Binding Authoritative Statement of Principles for a Global Consensus on the Management, Conservation and Sustainable Development of All Types of Forests."

45. See Corell 1999 and Corell and Betsill 2001 for details of the global negotiations on desertification.

46. Streck 2001, 75. The three implementing agencies of GEF are the UNDP, the UNEP, and the World Bank. It supports policies and programs in developing countries to address biodiversity, climate change, persistent organic pollutants, desertification, international water management, and ozone depletion.

47. Brack, Calder, and Dolun 2001, 2.

48. GEF has also leveraged an additional $12 billion in cofinancing from other sources.

49. See, for example, Chatterjee and Finger 1994; Sachs 1993.

50. Chatterjee and Finger 1994, 115–117; Connelly and Smith 1999, 206.

51. Shiva 1993b; Lohmann 1993.

52. For example, see Sachs 1993; *The Ecologist*, 1993.

53. Osborn and Bigg 1998.

54. Chapter 5 discusses the structure of the WTO and its relationship to environmental issues more fully.

55. Wade 2001.

56. Tabb 2000.

57. Mehta 2003, 122.

58. See Johannesburg Summit 2002, at www.johannesburgsummit.org.

59. Mehta 2003, 122.

60. Johannesburg Declaration on Sustainable Development, at www.johannes-burgsummit.org/html/documents/summit_clocs/1009wssd_pol_declaration.htm. Point 12 states: "The deep fault line that divides human society between the rich and the poor and the ever-increasing gap between the developed and developing worlds pose a major threat to global prosperity, security and stability." Point 14 states: "Globalization has added a new dimension to [global environmental problems]. The rapid integration of markets, mobility of capital and significant increases in investment flows around the world have opened new challenges and opportunities for the pursuit of sustainable development. But the benefits and costs of globalization are unevenly distributed, with developing countries facing special difficulties in meeting this challenge."

61. UN, 2003.

62. Doran 2002, 2.

63. Burg 2003, 116–118.

64. Wapner 2003.

65. The literature on global environmental governance is expanding quickly. A few examples include Hempel 1996; Clapp 1998; Haas 1999; Lipschutz 1999; Conca 2000; Vogler 2000, 2003; Paterson, Humphreys, and Pettiford 2003; Newell 2003; Falkner 2003; Bretherton 2003; Jordan, Wurzel, and Zito 2003.

66. See, for example, Litfin 1998.

67. Principle 21 reads: "States have, in accordance with the Charter of the United Nations and the principles of international law, the sovereign right to exploit their own resources pursuant to their own environmental policies, and the responsibility to ensure that activities within their jurisdiction or control do not cause damage to the environment of other states or of areas beyond the limits of national jurisdiction." The Stockholm Declaration on the Human Environment, adopted June 16, 1972, is available at www.unep.org/Documents/Default.asp?DocumentID=97&ArticleID=1503.

68. See Myers and Tucker 1987.

69. Clapp 1997, 129.

70. Andresen 2001a.

71. Downie and Levy 2000.

72. Chasek 2000.

73. UNEP, 2001a.

74. Political scientist Stephen Krasner (1983, 2) defines international regimes as "sets of implicit or explicit principles, norms, rules, and decision-making procedures around which actors' expectations converge in a given area of international relations."

75. Jacobsen and Weiss 1995, 121.

76. Paterson 2000b; Kutting 2000.

77. There is a growing environmental literature on regime effectiveness. See Susskind 1994; Victor, Raustiala, and Skolnikoff 1998; Weiss and Jacobson

1998; Wettestad 1999; Young 1999, 2001, 2002; Kütting 2000; Miles et al. 2001; Mitchell 2002; Hovi, Sprinz, and Underdal 2003.

78. For research on the interplay of environmental institutions, see Rosendal 2001; Andersen 2002; Young 2002; Selin and VanDeveer 2003.

79. Stokke and Thommessen 2003.

80. Union of International Associations, 1999; available at www.uia.org/statistics/organizations/ytb199.php and at www.uia.org/statistics/organizations/stybv296.php.

81. Lipschutz (with Mayer) 1996 discusses civil society and global environmental governance.

82. Betsill and Corell 2001 provide a framework for assessing NGO influence in global environmental negotiations.

83. Princen 1994.

84. Clapp 2001.

85. Wapner 2002.

86. Wapner 1996, 322. For an analysis of the global politics of whaling, see Peterson 1992; Stoett 1997; Andresen 2000, 2001b.

87. There is now an extensive literature on the role of NGOs in global environmental politics. See, for example, Princen and Finger 1994; Wapner 1996; Kolk 1996; Humphreys 1996; Jasanoff 1997; Najam 1998; Keck and Sikkink 1998; Auer 1998; Newell 2000; Tamiotti and Finger 2001; Bryner 2001; Corell and Betsill 2001; Hochstetler 2002; Ford 2003.

88. For details on the global political economy of persistent organic pollutants and the 2001 Stockholm Convention, see Lallas 2000–2001; Schafer 2002; Selin and Eckley 2003; Downie and Fenge 2003; Clapp 2003; Yoder 2003.

89. This charter can be found at www.europeangreens.org/info/globalgreencharter.html.

90. Susskind 1994; Gleckman 1995.

91. Nash and Ehrenfeld 1996.

92. Sklair 2001; Finger and Kilcoyne 1997.

93. Barnet and Cavanagh 1994; Korten 1995.

Chapter 4

1. All figures are from World Bank World Development Indicators Online: www.worldbank.org/data.

2. UNDP, 2001, 9.

3. See, for example, Tidsell 1993; Tietenberg 1992.

4. Simon 1996, 54, 67.

5. The Kuznets curve in economics depicts the relationship between growth and inequality over time, as hypothesized by Simon Kuznets (1955). Kuznets argued

that in the early stages of economic growth, income inequality would initially worsen, but that over time, once it passed a turning point, it would improve as growth continued. Kuznets based this hypothesis on time-series data for England, Germany, and the United States.

6. Grossman and Krueger 1991, 1995; World Bank, 1992.

7. Pempel 1998, 46.

8. Schreurs 2002, 36.

9. Schreurs 2002, 47. Also see Broadbent 1998; McKean 1981.

10. Bhattarai and Hammig 2001.

11. World Bank, 1992, 41.

12. Grossman and Krueger 1995.

13. Thomas and Belt 1997.

14. Munasinghe 1999.

15. World Bank, 1992, 25–32.

16. Mink 1993, 10–12.

17. UNDP, 1998, 57; see also Scherr 2000.

18. Cleaver and Schreiber 1992.

19. Barber and Schweithelm 2000, 17.

20. Quoted in Keane 1998.

21. Mink 1993.

22. World Bank, 1994, 161–162.

23. World Bank, 1992, 30–31.

24. Roodman 1998.

25. Schmidheiny 1992; DeSimone and Popoff 1997.

26. See the OECD website on sustainable consumption: www.oecd.org/env/consumption.

27. Keohane and Levy 1996.

28. Waring 1999.

29. Ekins, Hillman, and Hutchinson 1992, 36; Daly 1996, 40–42.

30. See, for example, Repetto et al. 1989. This study accounts for deforestation, soil erosion, and depletion of oil reserves in the national income and product accounts. It then recalculates Indonesia's annual GDP growth rate from 7.1 to 4 percent.

31. Waring 1999.

32. Davidson 2000, 6.

33. Ophuls 1973, 112–113.

34. Wackernagel et al. (2002, 9266–9271) conclude that humanity passed the biosphere's regenerative capacity in the late 1970s, and has been steadily moving beyond it since the mid-1980s. Humanity's load now sits at 120 percent.

35. Ayres 1998, 190.

36. Chapter 5 of Daly 1996 presents his parting suggestions for the World Bank on his resignation.

37. For example, Stern, Common, and Barbier 1996; Ezzati, Singer, and Kammen 2001.

38. World Bank, 1992, 41; Ansuategi and Escapa 2002.

39. Arrow et al. 1995.

40. Cameron 1996; Hall 2002.

41. Stern, Common, and Barbier 1996, 1156; Suri and Chapman 1998.

42. Clapp 1998, 2001.

43. Tidsell 2001.

44. Stern, Common, and Barbier 1996.

45. Arrow et al. 1995; Tidsell 2001, 187.

46. Duraiappah 1998.

47. Reardon and Vosti 1995.

48. Mander and Goldsmith 1996.

49. *The Ecologist*, 1993, 95.

50. Shiva 1993c.

51. Leach and Mearns 1996.

52. Bryant and Bailey 1997, 159.

53. Ostrom 1990. Elinor Ostrom is a professor of political science at Indiana University. She is one of the most influential theorists of the relationship between institutional arrangements and the management of common-pool resources. For Ostrom's recent research on the commons, see Dolšak and Ostrom 2003.

54. Bryant and Bailey 1997, 162. Also see Hardin 1968.

55. See, for example, the case studies in Peluso and Watts 2001.

56. Shiva 1997; Miller 2001.

57. Shiva 1993c, 1997.

58. Richards 1985; Fairhead and Leach 1995.

59. Broad 1994.

60. Shiva 1997.

61. Taylor 1995.

62. Broad 1994.

63. *The Ecologist*, 1993, 93.

64. Sachs 1999, 168.

65. UNDP, 1998, 46.

66. UNDP, 1998, 46.

67. UNDP, 1998, 50.

68. Myers 1997, 34.

69. Wackernagel and Rees 1996.

70. WWF, 2002, 22–26.

71. Wilson 2002, 23. See also Wackernagel and Rees 1996.

72. Wade 2001.

73. Kates 2000.

74. Myers 1997, 34.

75. Daly 1993b, 27–28.

76. Mies and Shiva 1993, 278.

77. *The Ecologist*, 1993, 140.

78. Shiva 1994; Sen 1994.

79. Schor 1998; Robbins 2002.

80. Sachs 1999; Princen, Maniates, and Conca 2002.

81. Sachs 1999, 167.

82. MacNeill, Winsemius, and Yakushiji 1991; Dauvergne 1997b.

83. Princen 1997.

84. Clapp 2002a.

85. See the Basel Action Network website: www.ban.org.

86. *The Ecologist*, 1993; Mander and Goldsmith 1996.

87. O'Connor 1998, 191–196; Dalby 2004.

88. See Maniates 2001 for a critical analysis of the individualization of environmental responsibility.

89. See, for example, the Global Green Charter at www.greens.org.

Chapter 5

1. For an overview of the history of this debate, see Williams 1994, 2001; Esty 1994.

2. WTO statistics online: www.wto.org/english/res_e/statis_e/statis_e.htm.

3. WTO data; see www.wto.org.

4. World Bank World Development Indicators Online: www.worldbank.org/data.

5. WTO statistics online: www.wto.org/english/res_e/statis_e/statis_e.htm.

6. WTO statistics online: www.wto.org/english/res_e/statis_e/statis_e.htm.

7. Bhagwati 1993, 2004.

8. Neumayer 2001, 103.

9. WTO, 1999, 29.

10. WTO, 1999, 4.

11. Neumayer 2001, 104.

12. World Bank, 1992, 67.

13. Vogel 2000, 366.

14. Logsdon and Husted 2000.

15. Bhagwati 1993, 48.

16. Weinstein and Charnovitz 2001, 150.

17. Dauvergne 2001a.

18. Bhagwati 1993, 44.

19. WTO, 1999, 44.

20. Sachs 1999, 183–184. See also Princen, Maniates, and Conca 2002.

21. Daly and Cobb 1989, 213–218.

22. Daly 2002, 13.

23. Goldsmith 1997, 3; Casson 2000.

24. Marchak 1995; Emberson-Bain 1994; Gedicks 2001.

25. Karliner 1997.

26. Dauvergne 1997b.

27. Princen 2002.

28. Ruitenbeek and Cartier 1998, 5.

29. Dauvergne 1997a discusses the theoretical and practical hurdles that the global community would need to overcome to create a genuinely sustainable trading system for tropical timber.

30. Daly 1993a, 25.

31. Conca 2000, 485.

32. Neumayer 2001, 108.

33. Daly 1993a, 26.

34. Esty 1994.

35. Porter 1999.

36. Goldsmith 1997, 7.

37. Karliner 1997, 144; Korten 1995, 177.

38. Korten 1995, 178–179.

39. Daly 1996, 157.

40. International Forum on Globalization, 2002, 76–77.

41. Hough 1998, 4, 27, 41. Also see the articles in Downie and Fenge 2003, for more details.

42. UN, 1992, 19.

43. See, for example, Esty 1994.

44. Neumayer 2001, 71–72.

45. Charnovitz 1995.

46. Neumayer 2001, 115–116.

47. See the FSC website, at www.fscoax.org. Also see Lipschutz 2001.

48. Neumayer 2001, 153–155.

49. Elliot 1998, 214. See also Arden-Clarke 1992, 130–132.

50. Pearce and Warford 1993, 27.

51. Arden-Clarke 1992, 130–131.

52. The GATT did have dispute panels to make rulings in the case of trade disputes, but any country could veto the resolution of a dispute, such that the decisions the GATT made were often ignored. The WTO dispute panels cannot be vetoed, and it has the authority to monitor and enforce its decisions with trade sanctions.

53. This round of trade talks was launched nearly 2 years after the failed attempt to launch a "Seattle Round" of trade talks in Seattle in the fall of 1999. Massive protests at that meeting led to the failure to initiate those talks.

54. As of April 4, 2003; see the WTO website: www.wto.org.

55. See Neumayer 2001, 24–25.

56. Esty 1994, 49–51.

57. Esty 1994, 30–31.

58. Neumayer 2001, 134.

59. Vogel 2000, 354.

60. Neumayer 2001, 126–127.

61. For a discussion, see Perkins 1998.

62. DeSombre and Barkin 2002.

63. For more discussion, see Perkins 1998.

64. Each of these other agreements is explained in full on the WTO website: www.wto.org. Also see IISD and UNEP, 2000; Neumayer 2001.

65. Neumayer 2001, 30, 128.

66. In June 2003 these countries were joined in their complaint by Australia, Brazil, Chile, Colombia, India, Mexico, Peru, and New Zealand.

67. See Neumayer 2001, 132–133.

68. Vogel 2000, 359–360.

69. See the WTO website: www.wto.org/english/tratop_e/dda_e/dohaexplained_e.htm#environment.

70. Stilwell and Tarasofsky 2001, 7–8.

71. Stilwell and Tarasofsky 2001, 10–11.

72. Houseman and Zaelke 1995, 315–328.

73. IISD and UNEP, 2000, 62.

74. Johannesburg Plan of Implementation, paragraph 92, available at www.johannesburgsummit.org/html/documents/summit_docs/2309_planfinal. htm.

75. Weinstein and Charnovitz 2001.

76. Esty 2000.

77. Esty 2001, 125.

78. Esty 1994, 41; 2001, 125–126.

79. Biermann 2000, 2001.

80. Wallach and Sforza 1999; Shiva 1993a.

81. Conca 2000, 489–492.

82. Conca 2000, 487.

83. Goldsmith 1997, 7.

84. Hines 2000, 130.

85. Hines 2000, 131.

86. Raynolds 2000, 298.

87. IISD and UNEP, 2000, 75.

88. See Hufbauer et al. 2000; IISD and UNEP, 2000.

89. Hufbauer et al. 2000.

90. On this debate, see Hogenboom 1998.

91. Hufbauer et al. 2000, 50.

92. Logsdon and Husted 2000, 373–374.

93. Vogel 2000, 363.

94. Hufbauer et al. 2000, 50.

95. Jacott, Reed, and Winfield 2001, 44, 52.

96. Prior to May 1, 2004, the membership of the European Union was as follows: Austria, Belgium, Denmark, Finland, France, Germany, Greece, Ireland, Italy, Luxembourg, the Netherlands, Portugal, Spain, Sweden, and the United Kingdom. On May 1, 2004, the following joined the EU: Cyprus, Czech Republic, Estonia, Hungary Latvia, Lithuania, Malta, Poland, Slovakia, and Slovenia.

97. See Geradin 2002.

98. Stevis and Mumme 2000, 31.

99. See the European Union's webpage on environmental activities, at www.europa.eu.int/pol/env/index_en.htm.

100. Geradin 2002, 128–129.

101. Zarsky 2002.

102. APEC's member economies include Australia, Brunei Darussalam, Canada, Chile, People's Republic of China, Hong Kong (China), Indonesia, Japan, Republic of Korea, Malaysia, Mexico, New Zealand, Papua New Guinea,

Peru, Philippines, Russia, Singapore, Chinese Taipei, Thailand, United States, and Vietnam.

103. Zarsky 2002.

Chapter 6

1. Forbes ranked all of these firms in the top twenty-five in 2003 in terms of sales, profits, assets, and market value. General Electric was number 1, Citigroup number 2, and ExxonMobil number 3. Wal-Mart Stores was number 1 in terms of sales, followed by General Motors. See www.forbes.com.

2. In 2001 China was the largest developing-country recipient of foreign direct investment, receiving some US$47 billion in investments (UNCTAD, 2002, 55).

3. Held et al. 1999, 191.

4. Willetts 2001, 362–363.

5. UNCTAD, 2001, 9; 2002, xv, 272.

6. French 2000, 16; World Bank, World Development Indicators Online, 2003: www.worldbank.org/data.

7. UNCTAD, 2001, 1; World Bank, World Development Indicators Online, 2003. Why, you might ask, the sudden drop in net inflows of FDI in 2001? UNCTAD (2002, xvi) notes: "After the record high levels of 2000, global flows declined sharply in 2001—for the first time in a decade. This was mainly the result of the weakening of the global economy, notably in the world's three largest economies which all fell into recession, and a consequent drop in the value of cross-border M and As [merger and acquisitions]." UNCTAD (2002) predicts later in the report that FDI will continue to rise in the future.

8. Scholte 2000, 82; UNCTAD, 2001, 9; UNCTAD, 2002, 310.

9. These figures are from UNCTAD, 2002, 7–9.

10. UNCTAD, 2002, 5.

11. UNCTAD, 2002, 7.

12. UNCTAD, 2002, 11.

13. UNCTAD, 2001, 2.

14. Anderson and Cavanagh 1999, 3.

15. Scholte 2000, 129.

16. Thompson and Strohm 1996.

17. Wheeler 2002, 1.

18. Porter 1999.

19. Low and Yeats 1992.

20. Leonard 1988; Pearson 1987; Low 1993.

21. Ferrantino 1997, 52.

22. Neumayer 2001, 55–56.

23. Wheeler 2001.

24. Mani and Wheeler 1998, 215.

25. Repetto 1994, 23.

26. Bhagwati 1993, 44; WTO, 1999, 44.

27. Internal World Bank memo, December 12, 1991, under the signature of Larry H. Summers. This leaked memo created a furor within the environmental community. Summers defended it, claiming it was intentionally ironic and provocative. In 1998 former World Bank economist Lant Pritchett told John Cassidy of the *New Yorker* that he was the actual author of the memo. Summers read, signed, and authorized its distribution within the World Bank, and therefore took responsibility for the contents. Pritchett (now a lecturer in public policy at Harvard University) claims the "leaked memo" was a condensed version of his longer memo. He feels the leak was intended to discredit and embarrass Summers.

28. GEMI, 1999, 6, 12.

29. World Bank, 1992, 67; Ferrantino 1997, 55–56.

30. Kent 1991, 10.

31. World Bank, 1999, 114.

32. World Bank, 1999, 130–135.

33. UN, 1992, 226.

34. Castleman 1985; UN Transnational Corporations and Management Division, 1992, 223.

35. ESCAP/UNCTC, 1990, 61; UNCTAD, 1993, 60.

36. See Rajan 2001.

37. Gladwin 1987; MacKenzie 2002.

38. MacKenzie 2002.

39. Morehouse 1994, 164.

40. Morehouse 1994, 167.

41. MacKenzie 2002.

42. Williams 1996, 777–779.

43. Sklair 1993, 79–80.

44. Molina 1993, 232.

45. Frey 2003; Bryant and Bailey 1997, 109; Korten 1995, 129.

46. See also Karliner 1997; Frey 1998; Castleman 1985; Ofreneo 1993.

47. Pearson 1987, 121; O'Neill 2001.

48. Hall 2002.

49. Clapp 2002b, 14–15.

50. Porter 1999, 136.

51. Neumayer 2001, 70–71.

52. Wheeler 2002, 7.

53. For logging see Dauvergne 2001a; Filer 1997; Marchak 1995. For mining, see Jackson and Banks 2003; Emberson-Bain 1994; Banks 1993. For industrial waste, see Clapp 2001. For oil, see Gedicks 2001.

54. See, for example, Global Witness, 1999, 2000; Sizer 1995; WRI, 1997; Rainforest Action Network at www.ran.org; Project Underground at www.moles.org.

55. Broad (with Cavanagh) (1993) title their book *Plundering Paradise*; Shiva (1997) titles her book *Biopiracy: The Plunder of Nature and Knowledge*.

56. Tokar 1997.

57. See Korten 1995. Korten (1999) calls for a "Post-Corporate World."

58. Karliner 1997.

59. See, for example, Dauvergne 2001a; Filer 1997; Marchak 1995.

60. For example, Potter 1993; Arentz 1996.

61. Japan imported over half of total log-production at the height of the log-export booms in the Philippines (1964–1973), Indonesia (1970–1980), and Sabah (1972–1987) (Dauvergne 1997b, 2).

62. WRI, 1997. Frontier forests are areas large and pristine enough to still retain full biodiversity.

63. Dauvergne 2001a.

64. Gedicks 2001, 29.

65. See Hutchinson 1998.

66. Gedicks 2001; Magno 2003; see also Miningwatch at www.miningwatch.ca; Project Underground at www.moles.org.

67. Kennedy (with Chatterjee and Moody) 1998; Emberson-Bain 1994.

68. This figure is from the late 1990s. See Project Underground, 1998.

69. Jermyn 2002.

70. Mitchell 1994 provides an in-depth analysis of treaty compliance and oil pollution at sea.

71. See Project Underground, 1998.

72. See, for example, Obi 1997; Rowell 1996.

73. Clapp 2001, 116–117.

74. "Greenpeace Exposes 'Corporate Criminal' Dow Chemical in South Africa," 2002.

75. Karliner 1994, 61.

76. See Greenpeace International, 2002.

77. Leighton, Roht-Arriaza, and Zarsky 2002, 96–98.

78. See Basel Action Network and Silicon Valley Toxics Coalition, 2002.

79. Prakash 2000.

80. Schmidheiny 1992.

81. Sklair 2001, 204.

82. Nash and Ehrenfeld 1996.

83. The UN Global Compact is available at www.unglobalcompact.org.

84. See Kollman and Prakash 2001.

85. DeSimone and Popoff 1997, 25.

86. Mol 2002.

87. Dryzek 1997, 141–143.

88. Prakash 2000, 3.

89. Levy and Newell 2000; Rowlands 2000.

90. Garcia-Johnson 2000.

91. Greer and Bruno 1997; Beder 1997.

92. Greer and Bruno 1997; Greenpeace Greenwash Detection kit, available at archive.greenpeace.org/~comms/97/summit/greenwash.html.

93. Rowell 1996.

94. Welford 1997.

95. Later revised and published as Greer and Bruno 1997.

96. Greer and Bruno 1997.

97. Sklair 2001, 200.

98. Bruno 2003.

99. CorpWatch, 2002.

100. Chatterjee and Finger 1994, 132.

101. Susskind 1992; Gleckman 1995.

102. Gleckman 1995, 95.

103. Chatterjee and Finger 1994.

104. Rutherford 2003, 14.

105. Corporate Europe Observer, 2001.

106. Sklair 2001; Levy and Newell 2002.

107. Newell and Paterson 1998; Levy 1997; Levy and Egan 1998; Levy and Newell 2002.

108. Roht-Arriaza 1995.

109. See Mann 2001; IISD and WWF, 2001.

110. See Mann 2001.

111. Neumayer 2001, 89.

112. IISD and UNEP, 2000, 58.

113. For an account, see Kobrin 1998; Roberts 1998; Clarke and Barlow 1997.

114. FOE England, Wales, and Northern Ireland, 1998.

115. OECD, 2000.

116. See FOE England, Wales, and Northern Ireland, 1998.

117. Leaver and Cavanagh 1996, 2.

118. Greenpeace International, 2002; FOEI, 2002.

Chapter 7

1. Hardin 1974.

2. World Bank, World Development Indicators Online: www.worldbank.org/data.

3. World Bank, World Development Indicators Online: www.worldbank.org/data.

4. These figures are in constant April 2001 U.S. dollars. See BIS, 2001.

5. Scholte 2000, 117.

6. In 2002 it provided an amount of development financing roughly equal to all of the regional development banks together (World Bank, 2003a, 126).

7. The World Bank Group today includes a number of agencies. In addition to the IBRD and the IDA, there are the International Finance Corporation (IFC), the Multilateral Investment Guarantee Agency (MIGA), and the International Center for the Settlement of Investment Disputes (ICSID).

8. World Bank website: www.worldbank.org.

9. World Bank, 2001a, 4.

10. Even today, however, individual researchers at the World Bank are still fully capable of publishing *classic* market-liberal arguments about the connections between the global political economy and global environmental change. The World Bank is a large and complex organization, and inevitably a range of views will "emerge" as "World Bank" views. Importantly, however, these all tend to fall within a range from market liberal to institutionalist.

11. Le Prestre 1989, 19; Wade 1997, 618. The name changed again in the early 1980s to the Office of Environmental and Scientific Affairs. For convenience, we call it the Office of the Environment.

12. See for example Rich 1994; Nelson 1995.

13. Wade 1997, 614.

14. Reed 1997, 229–230.

15. George 1988, 161; Wade 1997, 634–637.

16. Wade 1997, 637.

17. Rich 1994, 26–29.

18. Wade 1997, 688.

19. Rich 1994, 152–153.

20. Turaga 2000; Khagram 2000.

21. Turaga 2000, 246–248.

22. Wade 1997, 658–659.

23. Rich 1994, 125.

24. Fox and Brown 1998, 6.

25. Fox and Brown 1998, 508.

26. Wade 1997, 680.

27. Morse 1992.

28. Caufield 1996, 28.

29. Le Prestre 1989, 198–200; Reed 1997, 233.

30. Gutner 2002, 53.

31. Haas and Haas 1995, 268.

32. World Bank, 1992.

33. World Bank, 1995.

34. Fox 2000, 279.

35. Gutner 2002, 51.

36. Nelson 1995, 89–90.

37. Nelson 1995, 89–90.

38. FOE et al. 2001.

39. Durbin and Welch 2001.

40. FOEI, 1998.

41. Reed 1997, 236.

42. World Bank, 2001a, 20.

43. See the World Bank website at www.worldbank.org/environment/index.htm.

44. World Bank, 2001b.

45. Reed 1992; World Bank, 2001a, 61.

46. See World Bank, 1994; Pearce et al. 1995; Glover 1995.

47. World Bank, 2001a, 65.

48. World Bank, 2001a, 64.

49. Pearce et al. 1995, 55.

50. Glover 1995, 288.

51. World Bank, 2001a, 65.

52. Cheru 1992.

53. Hogg 1994; Horta 1991.

54. George 1992, 9–14.

55. Toye 1991, 192.

56. Rich 1994; Devlin and Yap 1994, 67–71.

57. George 1992, 3.

58. Hammond 1999.

59. FOEI, 1999, 10.

60. FOEI, 1999, 7.

61. See, for example, Keohane and Levy 1996.

62. UNEP, 2002b, 17.

63. Fairman and Ross 1996, 30–31.

64. International Forum on Globalization, 2002, 234.

65. Jordan 1994; Streck 2001.

66. Streck 2001, 73.

67. See the GEF website: www.gefweb.org.

68. Fairman 1996, 61.

69. Streck 2001, 75.

70. See the GEF website: www.gefweb.org/What_is_the_GEF/what_is_the_gef.html.

71. Rich 1994, 176–177.

72. Fairman 1996, 64.

73. Jordan 1994.

74. Chatterjee and Finger 1994, 153–154.

75. Streck 2001, 76. For an explanation of the voting procedures, see table 3.2 of this book.

76. Streck 2001, 86–87.

77. See Horta, Round, and Young 2002.

78. Horta, Round, and Young 2002, 15.

79. Young 2003.

80. This estimate relies on data from 1997 to 1999.

81. IDA, 2002.

82. Dauvergne 2001b. The top five recipients of Japanese ODA in 1995 were China (US$1,380 million), Indonesia (US$892 million), Thailand (US$667 million), India (US$506 million), and the Philippines (US$416 million).

83. Rich 2000, 34.

84. Goldzimer 2003, 4.

85. ECA Watch: www.eca-watch.org/eca/directory.html.

86. Rich 2000, 32.

87. UNEP webpage: www.uneptie.org/energy/act/fin/ECA; Goldzimer 2003, 2.

88. Goldzimer 2003, 4.

89. Goldzimer 2003, 3.

90. ECA Watch: www.eca-watch.org/eca/ecaflyer-english.pdf.

91. Rich 2000, 35.

92. Rich 2000, 35–36.

93. Berne Declaration et al., 1999.

94. See the OECD website for updates:
www.oecd.org/department/0,2688,en_2649_34181_1_1_1_1_1,00.html.

95. The Jakarta Declaration is available at www.eca-watch.org/goals/jakartadec.html.

96. Krugman 1998. Also see Dauvergne 1999.

97. Schmidheiny and Zorraquin 1996, 8–10.

98. Stephan Schmidheiny founded the World Business Council for Sustainable Development (WBCSD) in 1991. The WBCSD is "a coalition of 170 international companies united by a shared commitment to sustainable development via the three pillars of economic growth, ecological balance and social progress." See "About the WBCSD," available at www.wbcsd.ch.

99. See Schmidheiny and Zorraquin 1996.

100. See the Finance Initiatives website: www.unepfi.org.

101. Ganzi et al. 1998, 12.

102. This statement is available at www.unepfi.org/signatories/statements/fi/index.htm.

103. Quoted in Dennis Schultz, "Rivers of the Dammed," *The Australian Magazine*, posted by Muslimedia on May 16–31, 1997, at www.muslimedia.com/archives/sea98/dammed.htm.

104. Hardin 1974. For a radically different view see Singer 1977, who famously argues it is the moral duty of the rich to provide humanitarian assistance.

105. See the Jubilee UK website at www.jubilee2000uk.org.

106. See Oxfam et al., 2002.

Chapter 8

1. WBCSD, 2002, 2–3.

2. "Our Durable Planet," 1999.

3. WBCSD, 2002, three quotes from p. 3.

4. Lomborg 2001, 5.

5. DeSimone and Popoff 1997, 10–11.

6. See Kollman and Prakash 2001.

7. The quote is from the WBCSD website (www.wbcsd.org), under corporate responsibility. Also see Holme and Watts 2000; Watts and Holme 1999.

8. For example, see Chevron's CES program at www.chevron.com/about/pascagoula/env.

9. WBCSD 2002, 10.

10. These visions are not confined to market liberals. Authors like Jeremy Rifkin (2002, 8, 253) see an awe-inspiring future in hydrogen, a fuel that "never runs out" and "emits no carbon dioxide." It is, he claims, "a promissory note for humanity's future on Earth."

11. World Bank, 2003b, 3.

12. For more see WBCSD, 2000.

13. UN, 1992.

14. See Young 2002; Vogler 2003.

15. See, for example, the chapters in Victor, Raustiala, and Skolnikoff 1998; Young 1999.

16. See, for example, Mitchell 1994; Victor, Raustiala, and Skolnikoff 1998; Young 2001. See the notes to chapter 3 for a more complete list of the growing literature on environmental regimes.

17. UNEP, 2002b, 405.

18. Fox and Brown 1998.

19. Neumayer 2001.

20. Biermann 2000, 2001. See also Whalley and Zissimos 2001.

21. See Biermann 2001 on these various models.

22. See, for example, von Moltke 2001; Newell 2001; Najam 2003.

23. VanDeveer and Dabelko 2001.

24. UNEP, 2002a, 10.

25. World Bank, 2003b.

26. Rees 2002.

27. Ehrlich 1968.

28. Ehrlich 1968, 166.

29. Hardin 1968, 1974.

30. Daly 1993b, 335–340.

31. Brown, Gardner, and Halweil 1999, 18.

32. WWF, 2002, 20.

33. This is a popular theme of both moderate bioenvironmentalists and social greens. One of the best books on "branding" is Klein 2000.

34. Ryan and Durning 1997; Wackernagel and Rees 1996.

35. Daly 1993b, 325.

36. Daly 1996, 31–32.

37. See Daly and Cobb 1989 (the index was prepared with the assistance of Clifford Cobb).

38. See Cobb, Glickman, and Cheslog 2001.

39. See the graphics at www.rprogress.org/projects/gpi.

40. See the Friends of the Earth webpage on measuring progress: www.foe.co.uk/campaigns/sustainable_development/progress.

41. WWF, 2002, 2–3.

42. WWF, 2002, 4.

43. Rees 2002, 265.

44. Princen 2001, 2002.

45. Maniates 2001, 50.

46. Greenpeace International, 2001, 7; International Forum on Globalization, 2002, 226.

47. Hines 2000, 130–137.

48. Greenpeace International, 2001, 11.

49. Hines 2000, 131.

50. International Forum on Globalization, 2002, 223.

51. International Forum on Globalization, 2002, 227.

52. International Forum on Globalization, 2002, 234.

53. New Economics Foundation, 2002, 9–10.

54. See, for example, the Halifax Initiative on the Tobin Tax: www.currencytax.org.

55. See Greenpeace International, 2002.

56. See the Friends of the Earth website on corporate accountability at www.foei.org/publications/corporates/accountability.html.

57. International Forum on Globalization, 2002, 237–238.

58. International Forum on Globalization, 2002, 109.

59. Hines 2000, 68–69.

60. Meeker-Lowry 1996; Helleiner 2002.

61. Norberg-Hodge 1996, 394.

62. International Forum on Globalization, 2002, 108; *The Ecologist*, 1993.

63. Hines 2000, 213.

64. Hildyard 1995, 160.

65. See the Sierra Club website at www.sierraclub.org.

66. The parallel with the logic of Indian Prime Minister Indira Ghandi during her speech at the 1972 Stockholm Conference is striking: "Thus, we see when it comes to the depletion of natural resources and environmental pollution," she eloquently declared, "the increase of one inhabitant in an affluent country at his level of living is equivalent to an increase of many Asians, Africans or Latin Americans at their current material levels of living." It is reasonable to assume that Zuckerman's reasoning on the need to restrict immigration would horrify Ghandi, nicely demonstrating how similar ecological arguments can serve very different political purposes.

67. Zuckerman 2004.

68. See the Sierra Club of Canada: Population Policy, available at www.sierraclub.ca/national/programs/sustainable-economy/international/population-policy.shmtl.

References

Adams, William. 1990. *Green Development: Environment and Sustainability in the Third World*. London: Routledge.

Akula, Vikram. 1995. Grassroots Environmental Resistance in India. In Bron Taylor, ed., *Ecological Resistance Movements: The Global Emergence of Radical and Popular Environmentalism*, 127–145. Albany, NY: SUNY Press.

Allison, D. B., K. R. Fontaine, J. E. Manson, J. Stevens, and T. B. VanItallie. 1999. Annual Deaths Attributable to Obesity in the United States. *Journal of the American Medical Association* 282 (16): 1530–1538.

Andersen, Regine. 2002. The Time Dimension in International Regime Interplay. *Global Environmental Politics* 2 (3): 98–117.

Anderson, Sarah, and John Cavanagh. 1999. *Top 200: The Rise of Corporate Global Power*. Washington, DC: Institute for Policy Studies.

Andresen, Steinar. 2000. The Whaling Regime. In Steinar Andresen, Tora Skodvin, Jörgen Wettestad, and Arild Underdal, eds., *Science and Politics in International Environmental Regimes: Between Integrity and Involvement*, 35–70. Manchester: Manchester University Press.

Andresen, Steinar. 2001a. Global Environmental Governance: UN Fragmentation and Co-ordination. In *Yearbook of International Co-operation on Environment and Development*. London: Earthscan.

Andresen, Steinar. 2001b. The Whaling Regime: "Good" Policy But "Bad" Institutions? In Robert L. Friedheim, ed., *Towards a Sustainable Whaling Regime?*, 235–265. Seattle: University of Washington Press.

Ansuategi, Alberto, and Marta Escapa. 2002. Economic Growth and Greenhouse Gas Emissions. *Ecological Economics* 40 (1): 23–37.

Arden-Clarke, Charles. 1992. South-North Terms of Trade: Environmental Protection and Sustainable Development. *International Environmental Affairs* 4 (2): 122–139.

Arentz, Frans. 1996. Forestry and Politics in Sarawak: The Experience of the Penan. In Richard Howitt, with John Connell and Philip Hirsch, eds., *Resources, Nations and Indigenous Peoples: Case Studies from Australasia, Melanesia and Southeast Asia*, 202–211. Melbourne: Oxford University Press.

Arrow, Kenneth, Bert Bolin, Robert Costanza, Partha Dasgupta, Carl Folk, C. S. Holling, Bengt-Owe Jansson, Simon Levin, Karl-Göran Mäler, Charles Perrings, and David Pimentel. 1995. Economic Growth, Carrying Capacity and the Environment. *Science* 268 (April): 520–521.

Auer, Matthew. 1998. Colleagues or Combatants? Experts as Environmental Diplomats. *International Negotiation* 3 (2): 267–287.

Ayres, Robert. 1998. Eco-thermodynamics: Economics and the Second Law. *Ecological Economics* 26 (2): 189–209.

Bank for International Settlements (BIS). 2001. *Central Bank Survey of Foreign Exchange and Derivatives Market Activity in April 2001: Preliminary Global Data.* Press release, October 9, 2001. Available at www.bis.org/press/p011009.pdf.

Banks, Glenn. 1993. Mining Multinationals and Developing Countries: Theory and Practice in Papua New Guinea. *Applied Geography* 13: 313–327.

Barber, Charles Victor, and James Schweithelm. 2000. *Trial by Fire: Forest Fires and Forestry Policy in Indonesia's Era of Crisis and Reform.* Washington, DC: World Resources Institute.

Barlow, Maude, and Tony Clarke. 2002. *Global Showdown: How the New Activists Are Fighting Global Corporate Rule.* Rev. ed. Toronto: Stoddart.

Barnet, Richard J., and John Cavanagh. 1994. *Global Dreams: Imperial Corporations and the New World Order.* New York: Simon & Schuster.

Basel Action Network and Silicon Valley Toxics Coalition. 2002. *Exporting Harm: The High-Tech Trashing of Asia.* Available at www.svtc.org/cleancc/pubs/technotrash.pdf.

Beder, Sharon. 1997. *Global Spin: The Corporate Assault on Environmentalism.* Melbourne: Scribe.

Berne Declaration, Bioforum, Center for International Environmental Law, Environmental Defense, Eurodad, Friends of the Earth, Pacific Environment and Resources Center, and Urgewald. 1999. *A Race to the Bottom: Creating Risk, Generating Debt, and Guaranteeing Environmental Destruction.* Available at www.eca-watch.org/eca/race_bottom.pdf.

Bernstein, Steven. 2001. *The Compromise of Liberal Environmentalism.* New York: Columbia University Press.

Betsill, Michele M., and Elisabeth Corell. 2001. NGO Influence in International Environmental Negotiations: A Framework for Analysis. *Global Environmental Politics* 1 (3): 65–85.

Bhagwati, Jagdish. 1993. The Case for Free Trade. *Scientific American* 269 (5): 42–49.

Bhagwati, Jagdish. 2002. Coping with Antiglobalization: A Trilogy of Discontents. *Foreign Affairs* 81 (1): 2–7.

Bhagwati, Jagdish. 2004. In *Defense of Globalization.* Oxford: Oxford University Press.

Bhattarai, Madhusudan, and Michael Hammig. 2001. Institutions and the Environmental Kuznets Curve for Deforestation: A Cross-Country Analysis for Latin America, Africa and Asia. *World Development* 29 (6): 995–1010.

Biermann, Frank. 2000. The Case for a World Environment Organization. *Environment* 42 (9): 22–31.

Biermann, Frank. 2001. The Emerging Debate on the Need for a World Environment Organization. *Global Environmental Politics* 1 (1): 45–55.

Brac de la Perrière, Robert Ali, and Frank Seuret. 2000. *Brave New Seeds.* London: Zed Books.

Brack, Duncan, Fanny Calder, and Müge Dolun. 2001. *From Rio to Johannesburg: The Earth Summit and Rio + 10.* London: Royal Institute of International Affairs Briefing Paper, New Series No. 19.

Brenton, Tony. 1994. *The Greening of Machiavelli: The Evolution of International Environmental Politics.* London: Royal Institute of International Affairs/ Earthscan.

Bretherton, Charlotte. 2003. Movements, Networks, Hierarchies: A Gender Perspective on Global Environmental Governance. In David Humphreys, Matthew Paterson, and Lloyd Pettiford, eds., *Global Environmental Governance for the Twenty-First Century: Theoretical Approaches and Normative Considerations.* Special issue of *Global Environmental Politics* 3 (2): 103–119.

Broad, Robin. 1994. The Poor and the Environment: Friends or Foes? *World Development* 22 (6): 811–822.

Broad, Robin (with John Cavanagh). 1993. *Plundering Paradise: The Struggle for the Environment in the Philippines.* Berkeley: University of California Press.

Broadbent, Jeffrey. 1998. *Environmental Politics in Japan: Networks of Power and Protest.* Cambridge: Cambridge University Press.

Brown, Lester. 2003. *Plan B: Rescuing a Planet under Stress and a Civilization in Trouble.* New York: Norton.

Brown, Lester, Gary Gardner, and Brian Halweil. 1999. *Beyond Malthus: Nineteen Dimensions of the Population Challenge.* London: Norton.

Bruno, Kenny. 2003. *Presentation to the Public Eye on Davos.* Corporate PR Strategies Panel. Available at www.evb.ch/cm_data/panel prkbruno.pdf.

Bryant, Raymond. 1997. *The Political Ecology of Forestry in Burma, 1824–1994.* London: Hurst & Co.

Bryant, Raymond, and Sinéad Bailey. 1997. *Third World Political Ecology.* London: Routledge.

Bryner, Gary C. 2001. *Gaia's Wager: Environmental Movements and the Challenge of Sustainability.* Lanham, MD: Rowman and Littlefield.

Burg, Jericho. 2003. The World Summit on Sustainable Development: Empty Talk or Call to Action? *Journal of Environment and Development* 12 (1): 111–120.

Burtless, Gary, Robert Z. Lawrence, Robert E. Litan, and Robert J. Shapiro. 1998. *Globaphobia: Confronting Fears about Open Trade*. Washington, DC: Brookings Institution Press.

Caldwell, Lynton K. 1996. *International Environmental Policy: From the Twentieth to the Twenty-First Century*. Durham, NC: Duke University Press.

Cameron, Owen. 1996. Japan and South-East Asia's Environment. In Michael J. G. Parnell and Raymond L. Bryant, eds., *Environmental Change in South-East Asia: People, Politics and Sustainable Development*, 67–94. London: Routledge.

Carson, Rachel. 1962. *Silent Spring*. Boston: Houghton Mifflin.

Casson, Anne. 2000. *The Hesitant Boom: Indonesia's Oil Palm Sub-sector in an Era of Economic Crisis and Political Change*. Occasional Paper No. 29. Jakarta: Center for International Forestry Research.

Castleman, Barry. 1985. Double Standards in Industrial Hazards. In Jane Ives, ed., *The Export of Hazard*, 60–89. London: Routledge.

Caufield, Catherine. 1996. *Masters of Illusion: The World Bank and the Poverty of Nations*. London: Macmillan.

Charnovitz, Steve. 1995. Improving Environmental and Trade Governance. *International Environmental Affairs* 7 (1): 59–91.

Chasek, Pamela. 2000. The UN Commission on Sustainable Development: The First Five Years. In Pamela Chasek, ed., *The Global Environment in the Twenty-First Century: Prospects for International Cooperation*, 378–398. New York: United Nations University Press.

Chatterjee, Pratap, and Matthias Finger. 1994. *The Earth Brokers: Power, Politics and World Development*. London: Routledge.

Cheru, Fantu. 1992. Structural Adjustment, Primary Resource Trade and Sustainable Development in Sub-Saharan Africa. *World Development* 20 (4): 497–512.

Clapp, Jennifer. 1997. Environmental Threats in an Era of Globalization: An End to State Sovereignty? In Ted Schrecker, ed., *Surviving Globalism*, 123–140. New York: St. Martin's Press.

Clapp, Jennifer. 1998. The Privatization of Global Environmental Governance: ISO 14000 and the Developing World. *Global Governance* 4 (3): 295–316.

Clapp, Jennifer. 2001. *Toxic Exports: The Transfer of Hazardous Wastes from Rich to Poor Countries*. Ithaca, NY: Cornell University Press.

Clapp, Jennifer. 2002a. Distancing of Waste: Overconsumption in a Global Economy. In Tom Princen, Ken Conca, and Michael Maniates, eds., *Confronting Consumption*, 155–177. Cambridge, MA: MIT Press.

Clapp, Jennifer. 2002b. What the Pollution Havens Debate Overlooks. *Global Environmental Politics* 2 (2): 11–19.

Clapp, Jennifer. 2003. Transnational Corporate Interests and Global Environmental Governance: Negotiating Rules for Agricultural Biotechnology and Chemicals. *Environmental Politics* 12 (4): 1–23.

Clarke, Tony, and Maude Barlow. 1997. *MAI: The Multilateral Agreement on Investment and the Threat to Canadian Sovereignty.* Toronto: Stoddart.

Cleaver, Kevin, and Gotz Schreiber. 1992. Population, Agriculture and the Environment in Africa. *Finance and Development* 29 (2): 34–35.

Cobb, Clifford, Mark Glickman, and Craig Cheslog. 2001. *Genuine Progress Indicator 2000 Update.* Redefining Progress Issue Brief. Available at www.redefiningprogress.org/publications/2000_gpi_update.pdf.

Committee on World Food Security. 2001. *Assessment of the World Food Security Situation.* Available at www.fao.org/docrep/meeting/003/Y0147E/Y0147E00.htm#P79_3644.

Commoner, Barry. 1990. *Making Peace with the Planet.* New York: Pantheon Books.

Conca, Ken. 2000. The WTO and the Undermining of Global Environmental Governance. *Review of International Political Economy* 7 (3): 484–494.

Conca, Ken. 2001. Consumption and Environment in a Global Economy. *Global Environmental Politics* 1 (3): 56–57.

Conca, Ken, and Geoffrey D. Dabelko, eds. 1998. *Green Planet Blues: Environmental Politics from Stockholm to Kyoto.* 2nd ed. Boulder, CO: Westview Press.

Connelly, James, and Graham Smith. 1999. *Politics and the Environment: From Theory to Practice.* London: Routledge.

Corell, Elisabeth. 1999. Actor Influence in the 1993–97 Negotiations of the Convention to Combat Desertification. *International Negotiation* 4 (2): 197–223.

Corell, Elisabeth, and Michelle M. Betsill. 2001. A Comparative Look at NGO Influence in International Environmental Negotiations: Desertification and Climate Change. *Global Environmental Politics* 1 (4): 86–107.

Corporate Europe Observer. 2001. Industry's Rio + 10 Strategy: Banking on Feelgood PR. *CEO Quarterly Newsletter* 10 (December).

CorpWatch. 2002. *Greenwash + 10: The UN's Global Compact, Corporate Accountability and the Johannesburg Earth Summit.* Available at www.corpwatch.org/upload/document/gw10.pdf.

Costanza, Robert, John Cumberland, Herman Daly, Robert Goodland, and Richard Norgaard. 1997. *An Introduction to Ecological Economics.* Boca Raton, FL: St. Lucie Press.

Crosby, Alfred. 1986. *Ecological Imperialism: The Biological Expansion of Europe, 900–1900.* Cambridge: Cambridge University Press.

Dalby, Simon. 2004. Ecological Politics, Violence, and the Theme of Empire. *Global Environmental Politics* 4 (2): 1–11.

Daly, Herman. 1973. *Toward a Steady-State Economy.* San Francisco: W. H. Freeman.

Daly, Herman. 1993a. The Perils of Free Trade. *Scientific American* 269 (5): 50–57.

Daly, Herman. 1993b. The Steady-State Economy: Toward a Political Economy of Biophysical Equilibrium and Moral Growth. In Herman Daly and Kenneth Townsend, eds., *Valuing the Earth: Economics, Ecology, Ethics*, 324–356. Cambridge, MA: MIT Press.

Daly, Herman. 1996. *Beyond Growth: The Economics of Sustainable Development*. Boston: Beacon Press.

Daly, Herman. 1999. Globalization versus Internationalization—Some Implications. *Ecological Economics* 31 (1): 31–37.

Daly, Herman. 2002. Uneconomic Growth and Globalization in a Full World. *Natur and Kultur*. Available at www.puaf.umd.edu/faculty/papers/daly/unecon.pdf.

Daly, Herman, and John Cobb, Jr. 1989. *For the Common Good: Redirecting the Economy toward Community, the Environment, and a Sustainable Future*. Boston: Beacon Press.

D'Amato, Anthony, and Kirsten Engel. 1997. *International Environmental Law Anthology*. Cincinnati, OH: Anderson.

Dauvergne, Catherine. 2003. *Challenges to Sovereignty: Migration Laws for the 21st Century*. UN High Commission for Refugees New Issues in Refugee Research Working Paper Series, June.

Dauvergne, Peter. 1997a. A Model of Sustainable International Trade in Tropical Timber. *International Environmental Affairs* 9 (1): 3–21.

Dauvergne, Peter. 1997b. *Shadows in the Forest: Japan and the Politics of Timber in Southeast Asia*. Cambridge, MA: MIT Press.

Dauvergne, Peter. 1999. Asia's Environment After the 1997 Financial Meltdown: The Need of a Regional Response. *Asian Perspective: A Journal of Regional and International Affairs* 23 (3): 53–77.

Dauvergne, Peter. 2001a. *Loggers and Degradation in the Asia-Pacific: Corporations and Environmental Management*. Cambridge: Cambridge University Press.

Dauvergne, Peter. 2001b. The Rise of an Environmental Superpower? Evaluating Japanese Environmental Aid to Southeast Asia. In Javed Maswood, ed., *Japan and East Asian Regionalism*, 51–67. London: Routledge.

Davidson, Eric A. 2000. *You Can't Eat GNP: Economics as if Ecology Mattered*. Cambridge, MA: Perseus.

DeSimone, Livio, and Frank Popoff. 1997. *Eco-Efficiency: The Business Link to Sustainable Development*. Cambridge, MA: MIT Press.

DeSombre, Elizabeth R. 2002. *The Global Environment and World Politics*. London: Continuum.

DeSombre, Elizabeth R., and J. Samuel Barkin. 2002. Turtles and Trade: The WTO's Acceptance of Environmental Trade Restrictions. *Global Environmental Politics* 2 (1): 12–18.

Devlin, John, and Nonita Yap. 1994. Structural Adjustment Programmes and the UNCED Agenda: Explaining the Contradictions. In Caroline Thomas, ed., *Rio: Unravelling the Consequences*, 65–79. London: Frank Cass.

Dolšak, Nives, and Elinor Ostrom, eds. 2003. *The Commons in the New Millennium: Challenges and Adaptation.* Cambridge, MA: MIT Press.

Doran, Peter. 2002. *World Summit on Sustainable Development (Johannesburg) and Assessment for IISD.* IISD Briefing Paper, October 3. Available at www.iisd.org/pdf/2002/wssd_assessment.pdf.

Dornbusch, Rudiger, Stanley Fischer, and Gordon Sparks. 1993. *Marcroeconomics.* 4th Canadian ed. Toronto: McGraw-Hill Ryerson.

Downie, David, and Marc Levy. 2000. The UN Environment Programme at a Turning Point: Options for Change. In Pamela Chasek, ed., *The Global Environment in the Twenty-First Century: Prospects for International Cooperation,* 355–377. New York: UN University Press.

Downie, David Leonard, and Terry Fenge, eds. 2003. *Northern Lights against POPs: Combatting Toxic Threats in the Arctic.* Montreal/Kingston: McGill-Queen's University Press.

Dryzek, John. 1997. *The Politics of the Earth: Environmental Discourses.* Oxford: Oxford University Press.

Dryzek, John, and David Schlosberg, eds. 1998. *Debating the Earth: The Environmental Politics Reader.* Oxford: Oxford University Press.

Duraiappah, Anantha. 1998. Poverty and Environmental Degradation: A Review and Analysis of the Nexus. *World Development* 26 (12): 2169–2179.

Durbin, Andrea, and Carol Welch. 2001. Greening the Bretton Woods Institutions. *Foreign Policy in Focus* 5 (33). Available at www.foreignpolicy-infocus.org/briefs/vol5/v5n33bretton.html.

Easterbrook, Gregg. 1995. *A Moment on the Earth: The Coming Age of Environmental Optimism.* New York. Penguin.

Ecologist, The. 1993. *Whose Common Future? Reclaiming the Commons.* London: Earthscan.

Economic and Social Commission for the Asia Pacific (ESCAP) and the United Nations Center on Transnational Corporations (UNCTC). 1990. *Environmental Aspects of Transnational Corporation Activities in Pollution-Intensive Industries in Selected Asian and Pacific Developing Countries.* Bangkok: UN/ESCAP.

Ehrlich, Paul R. 1968. *The Population Bomb.* New York: Sierra Club–Ballantine.

Ehrlich, Paul R. 1981. An Economist in Wonderland. *Social Science Quarterly* 62 (1): 44–49.

Ehrlich, Paul R., and Anne H. Ehrlich. 1996. *Betrayal of Science and Reason: How Anti-Environmental Rhetoric Threatens Our Future.* Washington, DC: Island Press.

Ekins, Paul, Mayer Hillman, and Robert Hutchinson. 1992. *The Gaia Atlas of Green Economics.* New York: Anchor Books.

Elliot, Lorraine. 1994. *International Environmental Politics: Protecting the Antarctic.* London: Macmillan.

Elliot, Lorraine. 1998. *The Global Politics of the Environment*. New York: NYU Press.

Emberson-Bain, Àtu. 1994. De-Romancing the Stones: Gender, Environment and Mining in the Pacific. In Àtu Emberson-Bain, ed., *Sustainable Development or Malignant Growth? Perspectives of Pacific Island Women*, 91–110. Suva, Fiji: Marama Publications.

Esty, Daniel. 1994. *Greening the GATT: Trade, Environment and the Future*. Washington, DC: Institute for International Economics.

Esty, Daniel. 2000. Environment and the Trading System: Picking up the Post-Seattle Pieces. In J. J. Schott, ed., *The WTO After Seattle*, 243–253. Washington, DC: Institute for International Economics.

Esty, Daniel. 2001. Bridging the Trade-Environment Divide. *Journal of Economic Perspectives* 15 (3): 113–130.

Ezzati, Majid, Burton Singer, and Daniel Kammen. 2001. Towards an Integrated Framework for Development and Environment Policy: The Dynamics of Environmental Kuznets Curves. *World Development* 29 (8): 1421–1434.

Fairhead, James, and Melissa Leach. 1995. False Forest History, Complicit Social Analysis: Rethinking Some West African Environmental Narratives. *World Development* 23 (6): 1023–1053.

Fairman, David. 1996. The Global Environment Facility: Haunted by the Shadow of the Future. In Robert Keohane and Marc Levy, eds., *Institutions for the Earth*, 55–87. Cambridge, MA: MIT Press.

Fairman, David, and Michael Ross. 1996. Old Fads, New Lessons: Learning from Economic Development Assistance. In Robert Keohane and Marc Levy, eds., *Institutions for the Earth*, 29–51. Cambridge, MA: MIT Press.

Falkner, Robert. 2003. Private Environmental Governance and International Relations: Exploring the Links. In David Humphreys, Matthew Paterson, and Lloyd Pettiford, eds., *Global Environmental Governance for the Twenty-First Century: Theoretical Approaches and Normative Considerations*. Special issue of *Global Environmental Politics* 3 (2): 72–87.

Ferrantino, Michael. 1997. International Trade, Environmental Quality and Public Policy. *The World Economy* 20 (1): 43–72.

Filer, Colin, ed. 1997. *The Political Economy of Forest Management in Papua New Guinea*. London and Boroko: International Institute for Environment and Development and National Research Institute.

Finger, Matthias, and James Kilcoyne. 1997. Why Transnational Corporations Are Organizing to "Save the Global Environment." *The Ecologist* 27 (4): 138–142.

Finger, Matthias, and Ludivine Tamiotti. 1999. The Emerging Linkage between the WTO and the ISO: Implications for Developing Countries. Special issue on Globalisation and the Governance of the Environment, *IDS Bulletin* 30 (3): 8–16.

Ford, Lucy H. 2003. Challenging Global Environmental Governance: Social Agency and Global Civil Society. In David Humphreys, Matthew Paterson, and

Lloyd Pettiford, eds., *Global Environmental Governance for the Twenty-First Century: Theoretical Approaches and Normative Considerations*. Special issue of *Global Environmental Politics* 3 (2): 120–134.

Founex Report on Development and Environment. 1971. Available at www.southcentre.org/publications/conundrum/annex1.pdf.

Fox, Jonathan. 2000. The World Bank Inspection Panel: Lessons from the First Five Years. *Global Governance* 6 (3): 279–318.

Fox, Jonathan, and L. David Brown, eds. 1998. *The Struggle for Accountability: The World Bank, NGOs, and Grassroots Movements*. Cambridge, MA: MIT Press.

Frank, André Gunder. 1967. *Capitalism and Underdevelopment in Latin America*. New York: Monthly Review Press.

French, Hilary. 2000. *Vanishing Borders: Protecting the Planet in the Age of Globalization*. New York: Norton.

Frey, Scott. 1998. The Export of Hazardous Industries to the Peripheral Zones of the World System. *Journal of Developing Societies* 14 (1): 66–81.

Frey, Scott. 2003. The Transfer of Core-Based Hazardous Production Processes to the Export Processing Zones of the Periphery: The Maquiladora Centers of Northern Mexico. *Journal of World-Systems Research* 9 (2): 317–354.

Friedman, Thomas. 2002. Techno Logic. In *States of Discord*, A Debate between Thomas Friedman and Robert Kaplan. *Foreign Policy* 29 (March–April): 64–71.

Friends of the Earth (FOE), England, Wales and Northern Ireland. 1998. *A History of Attempts to Regulate the Activities of Transnational Corporations: What Lessons Can Be Learned?* Available at www.corporate-accountability.org/docs/FoE-US-paper-history_TNC-Regulation.doc.

Friends of the Earth (FOE), Environmental Defense, Sierra Club, International Rivers Network, and Rainforest Action Network. 2001. *Not in the Public Interest: The World Bank's Environmental Record*. Available at www.foe.org/res/pubs/pdf/wb.pdf.

Friends of the Earth International (FOEI). 1998. *Benchmarks for Mainstreaming the Environment: Environmental Reform Recommendations for the World Bank Group*. Available at www.foe.org/res/pubs/pdf/benchmarks.pdf.

Friends of the Earth International (FOEI). 1999. *The IMF: Selling the Environment Short*. Washington, DC: FOE. Available at www.foe.org/imf/index.html.

Friends of the Earth International (FOEI). 2002. *Towards Binding Corporate Accountability*. Available at www.foei.org/publications/corporates/accountability.html.

Ganzi, John, Frances Seymour, and Sandy Buffett (with Navroz Dubash). 1998. *Leverage for the Environment: A Guide to the Private Financial Services Industry*. Washington, DC: World Resources Institute.

Garcia-Johnson, Ronie. 2000. *Exporting Environmentalism: US Multinational Chemical Corporations in Brazil and Mexico*. Cambridge, MA: MIT Press.

Garrett, Geoffrey. 1998. *Partisan Politics in the Global Economy*. Cambridge: Cambridge University Press.

Gedicks, Al. 2001. *Resource Rebels: Native Challenges to Mining and Oil Corporations*. Boston: South End Press.

George, Susan. 1988. *A Fate Worse Than Debt*. London: Penguin.

George, Susan. 1992. *The Debt Boomerang*. London: Pluto.

Georgescu-Roegen, Nicholas. 1971. *The Entropy Law and the Economic Process*. Cambridge, MA: Harvard University Press.

Georgescu-Roegen, Nicholas. 1993. The Entropy Law and the Economic Problem. In Herman Daly and Kenneth Townsend, eds., *Valuing the Earth: Economics, Ecology, Ethics*, 37–49. Cambridge, MA: MIT Press.

Geradin, Damien. 2002. The European Community: Environmental Issues in an Integrated Market. In Richard Steinberg, ed., *The Greening of Trade Law: International Trade Organizations and Environmental Issues*, 117–154. Lanham, MD: Rowman and Littlefield.

Gladwin, Thomas. 1987. A Case Study of the Bhopal Tragedy. In Charles Pearson, ed., *Multinational Corporations, the Environment, and the Third World*, 223–239. Durham, NC: Duke University Press.

Gleckman, Harris. 1995. Transnational Corporations' Strategic Responses to "Sustainable Development." In Helge Ole Bergesen and Georg Parmann, eds., *Green Globe Yearbook*, 93–106. Oxford: Oxford University Press.

Gleick, Peter, William Burns, and Elizabeth Chalecki. 2002. *The World's Water 2002–2003: The Biennial Report on Freshwater Resources*. Washington, DC: Island Press.

Global Environment Facility (GEF). 2002. *GEF Replenishment*. Available at www.gefweb.org/Replenishment/replenishment.html.

Global Environment Management Initiative (GEMI). 1999. *Fostering Environmental Prosperity: Multinationals in Developing Countries*. Washington, DC: GEMI. Available at www.gemi.org/MNC_101.pdf.

Global Witness. 1999. *The Untouchables: Forest Crimes and Concessionaires—Can Cambodia Afford to Keep Them?* Briefing document, December. London: Global Witness.

Global, Witness. 2000. *Chainsaws Speak Louder Than Words*. Briefing document, May. London: Global Witness.

Glover, David. 1995. Structural Adjustment and the Environment. *Journal of International Development* 7 (2): 285–289.

Goldsmith, Edward. 1992. *The Way: An Ecological World-View*. London: Rider.

Goldsmith, Edward. 1997. Can the Environment Survive the Global Economy? *The Ecologist* 27 (6): 242–248.

Goldsmith, Edward, and Nicholas Hildyard. 1984. *The Social and Environmental Effects of Large Dams*. Wadebridge, UK: Wadebridge Ecological Centre.

Goldzimer, Aaron. 2003. Worse Than the World Bank? Export Credit Agencies—The Secret Engine of Globalization. *Food First Backgrounder* 9 (1). Available at www.foodfirst.org/pubs/backgrdrs/2003/w03v9n1.pdf.

Greenpeace Exposes "Corporate Criminal" Dow Chemical in South Africa. 2002. *World Finance Magazine*, August 29. Available at www.wfmagazine.com/News/wssd/greenpeaceexposescriminal.html.

Greenpeace International. 2001. *Safe Trade in the 21st Century: The Doha Edition*. Available at archive.greenpeace.org/politics/wto/doha_report.pdf.

Greenpeace International. 2002. *Corporate Crimes: The Need for an International Instrument on Corporate Accountability and Liability*. Amsterdam: Greenpeace International. Available at archive.greenpeace.org/earthsummit/docs/corpcrimes_1of3.pdf.

Greer, Jed, and Kenny Bruno. 1997. *Greenwash: The Reality Behind Corporate Environmentalism*. Penang: Third World Network; New York: Apex Press.

Grossman, Gene, and Alan Krueger. 1991. *Environmental Impact of a North American Free Trade Agreement*. Working Paper 3914. Cambridge, MA: National Bureau of Economic Research.

Grossman, Gene, and Alan Krueger. 1995. Economic Growth and the Environment. *Quarterly Journal of Economics* 110 (2): 353–377.

Grove, Richard. 1995. *Green Imperialism: Colonial Expansion, Tropical Island Edens and the Origins of Environmentalism, 1600–1860*. Cambridge: Cambridge University Press.

Grundmann, Reiner. 2001. *Transnational Environmental Policy: Reconstructing Ozone*. London: Routledge.

Gutner, Tamar L. 2002. *Banking on the Environment: Multilateral Development Banks and Their Environmental Performance in Central and Eastern Europe*. Cambridge, MA: MIT Press.

Haas, Peter. 1999. Social Constructivism and the Evolution of Multilateral Environmental Governance. In Aseem Prakash and Jeffrey Hart, eds., *Globalization and Governance*, 103–133. London: Routledge.

Haas, Peter, and Ernst B. Haas. 1995. Learning to Learn: Improving International Governance. *Global Governance* 1 (3): 255–285.

Haas, Peter, Robert Keohane, and Marc Levy, eds. 1993. *Institutions for the Earth: Sources of Effective International Environmental Protection*. Cambridge, MA: MIT Press.

Hall, Derek. 2002. Environmental Change, Protest and Havens of Environmental Degradation: Evidence from Asia. *Global Environmental Politics* 2 (2): 20–28.

Hammond, Ross. 1999. The Impact of IMF Structural Adjustment Policies on Tanzanian Agriculture. In the Development Gap and Friends of the Earth, eds., *The All-Too-Visible Hand: A Five-Country Look at the Long and Destructive Reach of the IMF*. Available at www.developmentgap.org/imftanzania.html.

Hardin, Garrett. 1968. The Tragedy of the Commons. *Science* 162: 1243–1248.

Hardin, Garrett. 1972. *Exploring New Ethics for Survival: The Voyage of the Spaceship Beagle.* Baltimore: Pelican.

Hardin, Garrett. 1974. Living on a Lifeboat. *Bioscience* 24 (10): 561–568.

Hardin, Garrett. 1985. *Filters against Folly: How to Survive Despite Economists, Ecologists, and the Merely Eloquent.* New York: Viking.

Hardin, Garrett. 1995. *Creative Altruism: An Ecologist Questions Motives.* Petoskey, MI: Social Contract Press.

Hardin, Garrett. 2000. *Living within Limits: Ecology, Economics, and Population Taboos.* Oxford: Oxford University Press.

Held, David, Anthony McGrew, David Goldblatt, and Jonathan Perraton. 1999. *Global Transformations: Politics, Economics, and Culture.* Stanford, CA: Stanford University Press.

Helleiner, Eric. 1994. *States and the Reemergence of Global Finance.* Ithaca, NY: Cornell University Press.

Helleiner, Eric. 2000. New Voices in the Globalization Debate: Green Perspectives on the World Economy. In Richard Stubbs and Geoffrey Underhill, eds., *Political Economy and the Changing Global Order*, 2nd ed., 60–69. Oxford: Oxford University Press.

Helleiner, Eric. 2002. Think Globally, Transact Locally: The Local Currency Movement and Green Political Economy. In Thomas Princen, Michael Maniates, and Ken Conca, eds., *Confronting Consumption*, 255–275. Cambridge, MA: MIT Press.

Helliwell, John F. 2002. *Globalization and Well-Being.* Vancouver, BC: UBC Press.

Hempel, Lamont C. 1996. *Environmental Governance: The Global Challenge.* Washington, DC: Island Press.

Hildyard, Nicholas. 1995. Liberation Ecology. In Helena Norberg-Hodge, Peter Goering, and Steven Gorelick, eds., *The Future of Progress: Reflections on Environment and Development*, 154–160. Foxhole, UK: Green Books.

Hill, Gladwin. 1969. Environment May Eclipse Vietnam as College Issue, *New York Times*, November 30: 1–2.

Hines, Colin. 2000. *Localization: A Global Manifesto.* London: Earthscan.

Hines, Colin. 2003. Time to Replace Globalization with Localization. *Global Environmental Politics* 3 (3): 1–7.

Hirst, Paul, and Grahame Thompson. 1999. *Globalization in Question: The International Economy and the Possibilities of Governance.* 2nd ed. Cambridge: Polity Press.

Hochstetler, Kathryn. 2002. After the Boomerang: Environmental Movements and Politics in the La Plata River Basin. *Global Environmental Politics* 2 (4): 35–57.

Hogenboom, Barbara. 1998. *Mexico and the NAFTA Environment Debate: The Transnational Politics of Economic Integration.* Utrecht, the Netherlands: International Books.

Hogg, Dominic. 1994. *The SAP in the Forest.* London: Friends of the Earth.

Holme, Richard, and Phil Watts. 2000. *Corporate Social Responsibility: Making Good Business Sense.* Geneva: World Business Council for Sustainable Development. Available at www.wbcsd.org/DocRoot/5mbU1sfWpqAgPpPpUqUe/csr2000.pdf.

Hoogvelt, Ankie. 1982. *The Third World in Global Development.* London: Macmillan.

Horta, Korinna. 1991. Multilateral Development Institutions Bear Major Responsibility for Solving Africa's Environmental Problems. In *Environmental Policies for Sustainable Growth in Africa.* Proceedings of the Fifth International Conference, Montclair State University, Upper Montelair, NJ. May 6: 21–28. Available at alpha.montclair.edu/~lebelp/CERAFConf1991.pdf.

Horta, Korinna, Robin Round, and Zoe Young. 2002. *The Global Environmental Facility: The First Ten Years—Growing Pains or Inherent Flaws?* Washington, D.C.: Environmental Defense and Halifax, NS: the Halifax Initiative, August. Available at www.environmentaldefense.org/documents/2265_First10YearsFinal.pdf.

Hough, Peter. 1998. *The Global Politics of Pesticides: Forging Consensus from Conflicting Interests.* London: Earthscan.

Houseman, Robert, and Durwood Zaelke. 1995. Mechanisms for Integration. In Robert Housman, Donald Goldberg, Brennan Van Dyke, and Durwood Zaelke, eds., *The Use of Trade Measures in Select Multilateral Environmental Agreements,* 315–328. Geneva: United Nations Environment Programme.

Hovi, Jon, Detlef F. Sprinz, and Arild Underdal. 2003. The Olso-Potsdam Solution to Measuring Regime Effectiveness: Critique, Response, and the Road Ahead. *Global Environmental Politics* 3 (3): 74–96.

Hufbauer, Gary, Daniel Esty, Diana Orejas, Luis Rubio, and Jeffrey Schott. 2000. *NAFTA and the Environment: Seven Years Later.* Washington, DC: Institute for International Economics.

Humphreys, David. 1996. Regime Theory and Non-Governmental Organizations: The Case of Forest Conservation. In David Potter, ed., *NGOs and Environmental Policies: Asia and Africa,* 90–115. London: Frank Cass.

Hurrell, Andrew, and Benedict Kingsbury. 1992. *The International Politics of the Environment.* Oxford: Oxford University Press.

Hutchinson, Moira. 1998. Beyond Best Practice: The Mining Sector. In Michelle Hibler and Rowena Beamish, eds., *Canadian Development Report 1998: Canadian Corporations and Social Responsibility,* 74–90. Ottawa: North South Institute.

Intergovernmental Panel on Climate Change (IPCC). 2001. *Climate Change 2001: The Scientific Basis.* Cambridge: Cambridge University Press.

International Development Association (IDA). 2002. *IDA, Grants and the Structure of Official Development Assistance.* Available at siteresources. worldbank.org/IDA/Resources/Seminar%20PDFs/grantsANDstructure.pdf.

International Forum on Globalization (IFG). 2002. *Alternatives to Economic Globalization: A Better World Is Possible.* San Francisco: Berrett-Koehler.

International Institute for Sustainable Development (IISD) and United Nations Environment Programme (UNEP). 2000. *Environment and Trade: A Handbook.* Winnipeg, MB: International Institute for Sustainable Development.

International Institute for Sustainable Development (IISD) and WWF Network. 2001. *Private Rights: Public Problems: A Guide to NAFTA's Controversial Chapter on Investor Rights.* Winnipeg, MB: International Institute for Sustainable Development.

International Monetary Fund (IMF). 2003. *IMF Quotas: A Factsheet.* Available at www.imf.org/external/np/exr/facts/quotas.htm.

International Organization for Standardization (ISO). 2003. *The ISO Survey Shows Continued Growth of ISO 9000 and ISO 14001 Certification.* Press release, 28 July. Available at www.iso.ch/iso/en/commcentre/pressreleases/2003/Ref864.html.

International Telecommunications Union (ITU). 2003. *Key Global Telecom Indicators for the World Telecommunication Service Sector.* Available at www.itu.int/ITU-D/ict/statistics/at_glance/KeyTelecom99.html.

International Union for Conservation of Nature and Natural Resources (IUCN). 1980. *World Conservation Strategy.* Geneva: IUCN, UNEP, WWF.

Jackson, Richard, and Glenn Banks. 2003. *In Search of the Serpent's Skin: The Story of the Porgera Gold Mine.* Brisbane: Boolorong Press.

Jacobson, Harold K., and Edith Brown Weiss. 1995. Strengthening Compliance with International Environmental Accords: Preliminary Observations from a Collaborative Project. *Global Governance* 1 (2): 119–148.

Jacott, Marisa, Cyrus Reed, and Mark Winfield. 2001. *The Generation and Management of Hazardous Wastes and Transboundary Hazardous Waste Shipments between Mexico, Canada, and the United States, 1990–2000.* Austin: Texas Center for Policy Studies. Available at www.texascenter.org/bordertrade/haznafta.htm.

Jasanoff, Sheila. 1997. NGOs and the Environment: From Knowledge to Action. *Third World Quarterly* 18 (3): 579–594.

Jasanoff, Sheila. 2001. Image and Imagination: The Formation of Global Environmental Consciousness. In Clark A. Miller and Paul N. Edwards, eds., *Changing the Atmosphere: Expert Knowledge and Environmental Governance,* 309–337. Cambridge: MIT Press.

Jermyn, Leslie. 2002. *Ecuadorian Oil, Debt and Poverty: A True Story.* Toronto: Global Aware Cooperative. August 19. Available at www.globalaware.org/Index_CD_eng.html.

Jordan, Andrew. 1994. Paying the Incremental Costs of Global Environmental Protection: The Evolving Role of GEF. *Environment* 36 (6): 13–20, 31–36.

Jordan, Andrew, Rüdiger K. W. Wurzel, and Anthony R. Zito. 2003. "New" Instruments of Environmental Governance: Patterns and Pathways of Change. *Environmental Politics* 12 (1): 3–24.

Joyner, Christopher C. 1998. *Governing the Frozen Commons: The Antarctic Regime and Environmental Protection*. Columbia: University of South Carolina Press.

Karliner, Joshua. 1994. The Environmental Industry. *The Ecologist* 24 (2): 59–63.

Karliner, Joshua. 1997. *The Corporate Planet, Ecology, and Politics in the Age of Globalization*. San Francisco: Sierra Club.

Kates, Robert. 2000. Population and Consumption: What We Know, What We Need to Know. *Environment* 42 (3): 10–19.

Keane, Fergal. 1998. Rain Forest Disaster Ignited by the Hand of Man. *Sunday Telegraph*, April 12.

Keck, Margaret, and Kathryn Sikkink. 1998. *Activists Beyond Borders: Advocacy Networks in International Politics*. Ithaca, NY: Cornell University Press.

Kennedy, Danny (with Pratap Chatterjee and Roger Moody). 1998. *Risky Business: The Grasberg Gold Mine*. Berkeley, CA: Project Underground. Available at www.moles.org/ProjectUnderground/downloads/risky_business.pdf.

Kent, Lawrence. 1991. *The Relationship between Small Enterprises and Environmental Degradation in the Developing World, with Special Emphasis on Asia*. Washington, DC: Development Alternatives, Inc.

Keohane, Robert, and Marc Levy, eds. 1996. *Institutions for Environmental Aid*. Cambridge, MA: MIT Press.

Khagram, Sanjeev. 2000. Toward Democratic Governance for Sustainable Development: Transnational Civil Society Organizing around Big Dams. In Ann Florini, ed., *The Third Force: The Rise of Transnational Civil Society*, 83–114. Washington, DC: Carnegie Endowment.

Kimerling, Judith. 2001. Corporate Ethics in the Era of Globalization: The Promise and Peril of International Environmental Standards. *Journal of Agricultural and Environmental Ethics* 14 (4): 425–455.

Klein, Naomi. 2000. *No Logo: Taking Aim at the Brand Bullies*. Toronto: Random House.

Kneen, Brewster. 1999. *Farmageddon*. Gabriola Island, BC: New Society.

Kobrin, Stephen. 1998. The MAI and the Clash of Globalizations. *Foreign Policy* 112 (Fall): 97–109.

Kolk, Ans. 1996. *Forests in International Environmental Politics: International Organisations, NGOs and the Brazilian Amazon*. Utrecht, the Netherlands: International Books.

Kollman, Kelly, and Aseem Prakash. 2001. Green by Choice? Cross National Variations in Firms' Responses to EMS-Based Environmental Regimes. *World Politics* 53 (3): 399–430.

Korten, David C. 1995. *When Corporations Rule the World.* West Hartford, CT: Kumarian Press; San Francisco: Berrett-Koehler.

Korten, David C. 1999. *The Post-Corporate World: Life After Capitalism.* West Hartford, CT: Kumarian Press; San Francisco: Berrett-Koehler.

Krasner, Stephen D. 1983. *International Regimes.* Ithaca, NY: Cornell University Press.

Krueger, Jonathan. 1999. *International Trade and the Basel Convention.* London: Earthscan.

Krugman, Paul. 1998. Saving Asia: It's Time to Get Radical. *Fortune* 135 (5): 74–80.

Krut, Riva, and Harris Gleckman. 1998. *ISO 14001: A Missed Opportunity for Global Sustainable Industrial Development.* London: Earthscan.

Kumar, Satish. 1996. Ghandi's Swadeshi: The Economics of Permanence. In Jerry Mander and Edward Goldsmith, eds., *The Case against the Global Economy and for a Turn Toward the Local,* 418–424. Sierra Club: San Francisco.

Kütting, Gabriela. 2000. *Environment, Society and International Relations: Towards More Effective International Environmental Agreements.* London: Routledge.

Kuznets, Simon. 1955. Economic Growth and Income Inequality. *American Economic Review* 45 (1): 1–28.

Laferrière, Eric. 2001. International Political Economy and the Environment: A Radical Ecological Perspective. In Dimitris Stevis and Valerie Assetto, eds., *The International Political Economy of the Environment: Critical Perspectives,* IPE Yearbook, vol. 12, 199–216. Boulder, CO: Lynne Reinner.

Lallas, Peter L. 2000–2001. The Role of Process and Participation in the Development of Effective International Environmental Agreements: A Study of the Global Treaty on Persistent Organic Pollutants. *UCLA Journal of Environmental Law and Policy* 19 (1): 83–152.

Layzer, Judith A. 2002. Science, Politics and International Environmental Policy. *Global Environmental Politics* 2 (4): 118–123.

Leach, Melissa, and Robin Mearns. 1996. *The Lie of the Land: Challenging Received Wisdom on the African Environment.* Westport, CT: Greenwood.

Lear, Linda. 1997. *Rachel Carson: The Life of the Author of Silent Spring.* New York: Holt.

Leaver, Erik, and John Cavanagh. 1996. Controlling Transnational Corporations. *Foreign Policy in Focus* 1 (6): 1–3.

Leighton, Michelle, Naomi Roht-Arriaza, and Lyuba Zarsky. 2002. *Beyond Good Deeds: Case Studies and a New Policy Agenda for Corporate Accountability.* Berkeley, CA: Nautilus Institute. Available at www.nautilus.org/cap/BeyondGoodDeedsCSRReportNautilusInstitute.pdf.

Leonard, H. Jeffrey. 1988. *Pollution and the Struggle for the World Product.* Cambridge: Cambridge University Press.

Le Prestre, Philippe. 1989. *The World Bank and the Environmental Challenge.* Selinsgrove, PA: Susquehanna University Press.

Levy, David. 1997. Business and International Environmental Treaties: Ozone Depletion and Climate Change. *California Management Review* 39 (3): 54–71.

Levy, David, and Daniel Egan. 1998. Capital Contests: National and Transnational Channels of Corporate Influence on the Climate Change Negotiations. *Politics and Society* 26 (3): 337–361.

Levy, David, and Peter Newell. 2000. Oceans Apart? Business Responses to Global Environmental Issues in Europe and the United States. *Environment* 42 (9): 8–20.

Levy, David, and Peter Newell. 2002. Business Strategy and International Environmental Governance: Toward a Neo-Gramscian Synthesis. *Global Environmental Politics* 2 (4): 84–101.

Lipschutz, Ronnie D. 1999. From Local Knowledge and Practice to Global Environmental Governance. In Martin Hewson and Timothy J. Sinclair, eds., *Approaches to Global Governance Theory,* 259–283. Albany, NY: SUNY Press.

Lipschutz, Ronnie D. 2001. Why Is There No International Forestry Law? An Examination of International Forestry Regulation, Both Public and Private. *UCLA Journal of Environmental Law and Policy* 19 (1): 155–182.

Lipschutz, Ronnie D. 2003. *Global Environmental Politics: Power, Perspectives, and Practice.* Washington, DC: Congressional Quarterly Press.

Lipschutz, Ronnie D. (with Judith Mayer). 1996. *Global Civil Society and Global Environmental Governance: The Politics of Nature from Place to Planet.* Albany, NY: SUNY Press.

Litfin, Karen, ed. 1998. *The Greening of Sovereignty in World Politics.* Cambridge, MA: MIT Press.

Logsdon, Jeanne, and Bryan Husted. 2000. Mexico's Environmental Performance under NAFTA: The First Five Years. *Journal of Environment and Development* 9 (4): 370–383.

Lohmann, Larry. 1993. Resisting Green Globalism. In Wolfgang Sachs, ed., *Global Ecology: A New Arena of Political Conflict,* 159–169. London: Zed Books.

Lomborg, Bjørn. 2001. *The Skeptical Environmentalist.* Cambridge: Cambridge University Press.

Lovelock, James. 1979. *Gaia: A New Look at Life on Earth.* Oxford: Oxford University Press.

Lovelock, James. 1995. *The Ages of Gaia: A Biography of Our Living Earth,* Rev. ed. New York: Norton.

Low, Patrick. 1993. The International Location of Polluting Industries and the Harmonization of Environmental Standards. In H. Munoz and R. Rosenberg, eds., *Difficult Liaison: Trade and the Environment in the Americas*, 21–50. London: Transaction.

Low, Patrick, and Alexander Yeats. 1992. Do "Dirty" Industries Migrate? World Bank Discussion Paper 159. In Patrick Low, ed., *International Trade and the Environment*, 89–104. Washington, DC: World Bank.

MacKenzie, Deborah. 2002. Fresh Evidence on Bhopal Disaster. *New Scientist* 176 (December 4): 6–7.

MacNeill, Jim, Pieter Winsemius, and Taizo Yakushiji. 1991. *Beyond Interdependence: The Meshing of the World's Economy and the Earth's Ecology*. New York: Oxford University Press.

Maddox, John. 1972. *The Doomsday Syndrome*. London: Macmillan.

Magno, Francisco. 2003. Human and Ecological Security: The Anatomy of Mining Disputes in the Philippines. In David Dewitt and Carolina Hernandez, eds., *Development and Security in Southeast Asia, Volume 1: The Environment*, 115–136. Aldershot, UK: Ashgate.

Malthus, Thomas Robert. 1798 (1st ed.); 1826 (6th ed.). *Essay on the Principle of Population*. Available at www.econlib.org/library/Malthus/malPlong.html.

Mander, Jerry, and Edward Goldsmith, eds. 1996. *The Case against the Global Economy: and for a Turn toward the Local*. San Francisco: Sierra Club.

Mani, Muthukumara, and David Wheeler. 1998. In Search of Pollution Havens? Dirty Industry in the World Economy, 1960–1993. *Journal of Environment and Development* 7 (3): 215–247.

Maniates, Michael F. 2001. Individualization: Plant a Tree, Buy a Bike, Save the World? *Global Environmental Politics* 1 (3): 31–52.

Maniates, Michael F., ed. 2003. *Encountering Global Environmental Politics: Teaching, Learning, and Empowering Knowledge*. Lanham, MD: Rowman and Littlefield.

Mann, Howard. 2001. *Private Rights, Public Problems: A Guide to NAFTA's Chapter on Investor Rights*. Winnipeg: IISD. Available at www.iisd.org/pdf/trade_citizensguide.pdf.

Marchak, M. Patricia. 1995. *Logging the Globe*. Montreal and Kingston: McGill-Queen's University Press.

Marland, G., T. A. Boden, and R. J. Andres. 2001. Global, Regional, and National CO_2 Emissions. In Carbon Dioxide Information Analysis Center, ed., *Trends: A Compendium of Data on Global Change*. Oak Ridge, TN: Carbon Dioxide Information Analysis Center, Oak Ridge National Laboratory, U.S. Department of Energy.

McKean, Margaret A. 1981. *Environmental Protest and Citizen Politics in Japan*. Berkeley: University of California Press.

McMurtry, John. 1999. *The Cancer Stage of Capitalism*. London: Pluto.

Meadows, Donella H. 1993. The Limits to Growth Revisited. In Paul Ehrlich and John P. Holdren, eds., *The Cassandra Conference: Resources and the Human*

Predicament, 257–269. College Station: Texas A&M University Press. Excerpted in Mark Zacher, ed., *International Political Economy of Natural Resources*, vol. 2, Cheltenham, UK: Edward Elgar, 1993.

Meadows, Donella H., Dennis L. Meadows, William W. Behrens, and Jørgen Randers. 1972. *The Limits to Growth*. New York: Club of Rome.

Meadows, Donella H., Dennis L. Meadows, and Jørgen Randers. 1992. *Beyond the Limits: Confronting Global Collapse, Envisioning a Sustainable Future*. White River Junction, VT: Chelsea Green.

Meeker-Lowry, Susan. 1996. Community Money: The Potential of Local Currency. In Jerry Mander and Edward Goldsmith, eds., *The Case against the Global Economy and for a Turn Toward the Local*, 446–459. San Francisco: Sierra Club Books.

Mehta, Sandeep. 2003. The Johannesburg Summit from the Depths. *Journal of Environment and Development* 12 (1): 121–128.

Meyer-Abich, Klaus M. 1992. Winners and Losers in Climate Change. In Wolfgang Sachs, ed., *Global Ecology: A New Arena of Political Conflict*, 68–87. London: Zed Books.

Mies, Maria, and Vandana Shiva. 1993. *Ecofeminism*. London: Zed Books.

Miles, Edward L., Arild Underdal, Steinar Andresen, Jørgen Wettestad, Jon Birger Skjærseth, and Elaine M. Carlin. 2001. *Environmental Regime Effectiveness: Confronting Theory with Evidence*. Cambridge, MA: MIT Press.

Mill, John Stuart. 1848. *Principles of Political Economy*.

Miller, Marian. 2001. Tragedy for the Commons: The Enclosure and Commodification of Knowledge. In Dimitris Stevis and Valerie Assetto, eds., *The International Political Economy of the Environment*, 111–134. Boulder, CO: Lynne Reinner.

Mink, Stephen. 1993. *Poverty, Population and the Environment*. World Bank Discussion Paper 189. Washington, DC: World Bank.

Mitchell, Ronald. 1994. *Intentional Oil Pollution at Sea: Environmental Policy and Treaty Compliance*. Cambridge, MA: MIT Press.

Mitchell, Ronald. 2002. A Quantitative Approach to Evaluating International Environmental Regimes. *Global Environmental Politics* 2 (4): 58–83.

Mol, Arthur. 2002. Ecological Modernization and the Global Economy. *Global Environmental Politics* 2 (2): 92–115.

Molina, David. 1993. A Comment on Whether Maquiladoras Are in Mexico for Low Wages or to Avoid Pollution Abatement Costs. *Journal of Environment and Development* 2 (1): 221–241.

Morehouse, Ward. 1994. Unfinished Business: Bhopal Ten Years After. *The Ecologist* 24 (5): 164–168.

Morse, Bradford. 1992. *Sardar Sarovar: Report of the Independent Review*. Ottawa: Resource Futures International.

Munasinghe, Mohan. 1999. Is Environmental Degradation an Inevitable Consequence of Economic Growth: Tunneling through the Environmental Kuznets Curve. *Ecological Economics* 29 (1): 89–109.

Murphy, Craig. 1983. What the Third World Wants: An Interpretation of the Development and Meaning of the New International Economic Order. *International Studies Quarterly* 27: 55–76.

Mushita, Andrew T., and Carol B. Thompson. 2002. Patenting Biodiversity? Rejecting WTO/TRIPS in Southern Africa. *Global Environmental Politics* 2 (1): 65–82.

Myers, Norman. 1979. *The Sinking Ark: A New Look at the Problem of Disappearing Species.* New York: Pergamon Press.

Myers, Norman. 1997. Consumption in Relation to Population, Environment and Development. *The Environmentalist* 17: 33–44.

Myers, Norman, and Richard P. Tucker. 1987. Deforestation in Central America: Spanish Legacy and North American Consumers. *Environmental Review* 11 (1): 55–71.

Najam, Adil. 1998. Searching for NGO Effectiveness. *Development Policy Review* 16 (3): 305–310.

Najam, Adil. 2003. The Case against a New International Environmental Organization. *Global Governance* 9 (3): 367–384.

Nash, Jennifer, and John Ehrenfeld. 1996. Code Green. *Environment* 37 (1): 16–45.

Nelson, Paul J. 1995. *The World Bank and NonGovernmental Organizations: The Limits of Apolitical Development.* London: Macmillan.

Neumayer, Eric. 2001. *Greening Trade and Investment: Environmental Protection without Protectionism.* London: Earthscan.

New Economics Foundation (NEF). 2002. *Chasing Shadows: Re-Imagining Finance for Development.* Available at www.jubileeresearch.org/analysis/reports/chasingshadows.pdf.

Newell, Peter, and Matthew Paterson. 1998. A Climate for Business: Global Warming, the State, and Capital. *Review of International Political Economy* 5 (4): 679–703.

Newell, Peter. 2000. *Climate for Change: Non-State Actors and the Global Politics of the Greenhouse.* Cambridge: Cambridge University Press.

Newell, Peter. 2001. New Environmental Architectures and the Search for Effectiveness. *Global Environmental Politics* 1 (1): 35–44.

Newell, Peter. 2003. Globalization and the Governance of Biotechnology. In David Humphreys, Matthew Paterson, and Lloyd Pettiford, eds., *Global Environmental Governance for the Twenty-First Century: Theoretical Approaches and Normative Considerations,* Special issue of *Global Environmental Politics* 3 (2): 56–71.

Norberg-Hodge, Helena. 1996. Shifting Direction: From Global Dependence to Local Interdependence. In Jerry Mander and Edward Goldsmith, eds., *The Case against the Global Economy and For a Turn toward the Local*, 393–406. San Francisco: Sierra Club Books.

Obi, Cyril. 1997. Globalization and Local Resistance: The Case of Ogoni versus Shell. *New Political Economy* 2 (1): 137–148.

O'Connor, James. 1998. *Natural Causes: Essays in Ecological Marxism.* New York: Guilford.

Ofreneo, Rene. 1993. Japan and the Environmental Degradation of the Philippines. In Michael Howard, ed., *Asia's Environmental Crisis*, 201–219. Boulder, CO: Westview Press.

O'Neill, Kate. 2000. *Waste Trading among Rich Nations: Building A New Theory of Environmental Regulation.* Cambridge, MA: MIT Press.

O'Neill, Kate. 2001. The Changing Nature of Global Waste Management for the 21st Century. *Global Environmental Politics* 1 (1): 77–98.

Ophuls, William. 1973. *Ecology and the Politics of Scarcity: Prologue to a Political Theory of the Steady State.* San Francisco: Freeman.

Organization for Economic Cooperation and Development (OECD). 2000. *The OECD Guidelines for Multinational Enterprises.* Available at www.oecd.org/dataoecd/56/36/1922428.pdf.

Osborn, Derek, and Tom Bigg. 1998. *Earth Summit II: Outcomes and Analysis.* London: Earthscan.

Ostrom, Elinor. 1990. *Governing the Commons: The Evolution of Institutions for Collective Action.* Cambridge: Cambridge University Press.

Our Durable Planet. 1999. *The Economist.* (Twentieth Century Survey) 352 (September 11): 29–31.

Oxfam, CAFOD, Christian Aid, and Eurodad. 2002. *A Joint Submission to the World Bank and IMF Review of HIPC and Debt Sustainability.* Available at www.oxfam.org.uk/what_you_can_do/campaign/educationnow/downloads/debt _sustainability_0802.doc.

Paarlberg, Robert. 2000. The Global Food Fight. *Foreign Affairs* 79 (3): 24–38.

Paehlke, Robert. 2003. *Democracy's Dilemma: Environment, Social Equity, and the Global Economy.* Cambridge, MA: MIT Press.

Parson, Edward A. 2002. *Protecting the Ozone Layer: Science, Strategy, and Negotiation in the Shaping of a Global Environmental Regime.* Oxford: Oxford University Press.

Paterson, Matthew. 1996. *Global Warming and Global Politics.* London: Routledge.

Paterson, Matthew. 2000a. Car Culture and Global Environmental Politics. *Review of International Studies* 26 (2): 253–270.

Paterson, Matthew. 2000b. *Understanding Global Environmental Politics: Domination, Accumulation, Resistance.* London: Palgrave Macmillan.

Paterson, Matthew. 2001a. Climate Policy as Accumulation Strategy: The Failure of COP 6 and Emerging Trends in Climate Politics. *Global Environmental Politics* 1 (2): 10–17.

Paterson, Matthew. 2001b. Risky Business: Insurance Companies in Global Warming Politics. *Global Environmental Politics* 1 (3): 18–42.

Paterson, Matthew, David Humphreys, and Lloyd Pettiford. 2003. Conceptualizing Global Environmental Governance: From Interstate Regimes to Counter-Hegemonic Struggles. In David Humphreys, Matthew Paterson, and Lloyd Pettiford, eds., *Global Environmental Governance for the Twenty-First Century: Theoretical Approaches and Normative Considerations*, Special issue of *Global Environmental Politics* 3 (2): 1–10.

Pearce, David, Neil Adger, David Maddison, and Dominic Moran. 1995. Debt and the Environment. *Scientific American* 272 (June): 52–56.

Pearce, David, and Jeremy Warford. 1993. *World without End: Economics, Environment and Sustainable Development*. New York: Oxford University Press.

Pearson, Charles. 1987. Environmental Standards, Industrial Relocation, and Pollution Havens. In Charles Pearson, ed., *Multinational Corporations, the Environment, and the Third World*, 113–128. Durham, NC: Duke University Press.

Pearson, Charles, ed. 1987. *Multinational Corporations, the Environment, and the Third World*. Durham, NC: Duke University Press.

Peluso, Nancy Lee, and Michael Watts, eds. 2001. *Violent Environments*. Ithaca, NY: Cornell University Press.

Pempel, T. J. 1998. *Regime Shift: Comparative Dynamics of the Japanese Political Economy*. Ithaca, NY: Cornell University Press.

Perkins, Nancy. 1998. *World Trade Organizations: United States—Import Prohibition of Certain Shrimp and Shrimp Products*. Available at American Society for International Law website: www.asil.org/ilm/nancy.htm.

Peterson, M. J. 1992. Whalers, Cetologists, Environmentalists and the International Management of Whaling. *International Organization* 46 (1): 147–186.

Pojman, Louis P. 2000. *Global Environmental Ethics*. Mountain View, CA: Mayfield.

Population Division of the Department of Economic and Social Affairs of the UN Secretariat. 2001. *World Population Prospects: The 2000 Revision* (Highlights). Available at www.un.org/esa/population/publications/wpp2000/highlights.pdf.

Porter, Gareth. 1999. Trade Competition and Pollution Standards: "Race to the Bottom" or "Stuck at the Bottom"? *Journal of Environment and Development* 8 (2): 133–151.

Porter, Gareth, Janet Welsh Brown, and Pamela S. Chasek. 2000. *Global Environmental Politics*. 3rd ed. Boulder, CO: Westview Press.

Potter, Lesley. 1993. The Onslaught on the Forests in South-East Asia. In Harold Brookfield and Yvonne Byron, eds., *South-East Asia's Environmental Future: The Search for Sustainability*, 103–123. Tokyo: United Nations University Press; Melbourne: Oxford University Press.

Pounds, J. Alan, and Robert Puschendorf. 2004. Clouded Futures. *Nature* 427 (January 8): 107–108.

Prakash, Aseem. 2000. *Greening the Firm: The Politics of Corporate Environmentalism*. Cambridge: Cambridge University Press.

Princen, Thomas. 1994. NGOs: Creating a Niche in Environmental Diplomacy. In Thomas Princen and Matthias Finger, eds., *Environmental NGOs in World Politics: Linking the Local and the Global*, 29–47. London: Routledge.

Princen, Thomas. 1997. The Shading and Distancing of Commerce: When Internalization Is Not Enough. *Ecological Economics* 20 (3): 235–253.

Princen, Thomas. 2001. Consumption and Its Externalities: Where Economy Meets Ecology. *Global Environmental Politics* 1 (3): 11–30.

Princen, Thomas. 2002. Distancing: Consumption and the Severing of Feedback. In Thomas Princen, Michael Maniates, and Ken Conca, eds., *Confronting Consumption*, 103–131. Cambridge, MA: MIT Press.

Princen, Thomas, and Matthias Finger, eds. 1994. *Environmental NGOs in World Politics: Linking the Local and the Global*. London: Routledge.

Princen, Thomas, Michael Maniates, and Ken Conca, eds. 2002. *Confronting Consumption*. Cambridge, MA: MIT Press.

Project Underground. 1998. *Drilling to the Ends of the Earth*. Available at moles.org/ProjectUnderground/motherlode/drilling/intro.html.

Rajan, S. Ravi. 2001. Toward a Metaphysic of Environmental Violence: The Case of the Bhopal Gas Disaster. In Nancy Lee Peluso and Michael Watts, eds., *Violent Environments*, 380–398. Ithaca, NY: Cornell University Press.

Rangan, Haripriya. 1996. From Chipko to Uttaranchal: Development, Environment and Social Protest in the Gharhwal Himalayas, India. In Richard Peet and Michael Watts, eds., *Liberation Ecologies: Environment, Development, Social Movements*, 205–226. New York: Routledge.

Rapley, John. 2002. *Understanding Development*. Boulder, CO: Lynne Reinner.

Raynolds, Laura. 2000. Re-embedding Global Agriculture: The International Organic and Fair Trade Movements. *Agriculture and Human Values* 17 (1): 297–309.

Reardon, Thomas, and Stephen Vosti. 1995. Links between Rural Poverty and the Environment in Developing Countries: Asset Categories and Investment Poverty. *World Development* 23 (9): 1495–1506.

Reed, David. 1997. The Environmental Legacy of the Bretton Woods: The World Bank. In Oran K. Young, ed., *Global Governance: Drawing Insights from the Environmental Experience*, 227–245. Cambridge, MA: MIT Press.

Reed, David, ed. 1992. *Structural Adjustment and the Environment*. Boulder, CO: Westview Press.

Rees, William E. 2002. Globalization and Sustainability: Conflict or Convergence? *Bulletin of Science, Technology and Society* 22 (4): 249–268.

Rees, William E., and Laura Westra. 2003. When Consumption Does Violence: Can There Be Sustainability and Environmental Justice in a Resource-Limited World? In Julian Agyeman, Robert Bullard, and Bob Evans, eds., *Just Sustainabilities: Development in an Unequal World*, 99–124. London: Earthscan.

Repetto, Robert. 1994. *Trade and Sustainable Development*. UNEP Environment and Trade Series No. 1. Geneva: United Nations Environment Programme.

Repetto, Robert, William McGrath, Michael Wells, Christine Beer, and Fabrizio Rossini. 1989. *Wasting Assets: Natural Resources in the National Income and Product Accounts*. Washington, DC: World Resources Institute.

Ricardo, David. 1817. *On the Principles of Political Economy and Taxation*. Available at www.econlib.org/library/Ricardo/ricP.html.

Rich, Bruce. 1994. *Mortgaging the Earth: The World Bank, Environmental Impoverishment, and the Crisis of Development*. London: Earthscan.

Rich, Bruce. 2000. Exporting Destruction. *Environmental Forum* (September–October): 32–40.

Richards, Paul. 1985. *Indigenous Agricultural Revolution*. London: Unwin Hyman.

Rifkin, Jeremy. 2002. *The Hydrogen Economy: The Creation of the Worldwide Energy Web and the Redistribution of Power on Earth*. New York: Tarcher/Putnam.

Robbins, Richard H. 2002. *Global Problems and the Culture of Capitalism*, 2nd ed. Boston: Allyn and Bacon.

Roberts, Joseph. 1998. Multilateral Agreement on Investment. *Monthly Review* 50 (5): 23–32.

Rodney, Walter. 1972. *How Europe Underdeveloped Africa*. Dar es Salaam: Tanzania Publishing House.

Rogers, Adam. 1993. *The Earth Summit: A Planetary Reckoning*. Los Angeles: Global View Press.

Roht-Arriaza, Naomi. 1995. Shifting the Point of Regulation: The International Organization for Standardization and Global Law-Making on Trade and the Environment. *Ecology Law Quarterly*, 22 (3): 479–539.

Roodman, David Malin. 1998. *The Natural Wealth of Nations: Harnessing the Market for the Environment*. New York: Norton.

Rosendal, G. Kristin. 2001. Impact of Overlapping International Regimes: The Case of Biodiversity. *Global Governance* 7 (2): 95–117.

Rowell, Andrew. 1996. *Green Backlash: Global Subversion of the Environment Movement*. London: Routledge.

Rowland, Wade. 1973. *The Plot to Save the World*. Toronto: Clarke, Irwin and Company.

Rowlands, Ian. 1995. *The Politics of Global Atmospheric Change*. Manchester: Manchester University Press; New York: St. Martin's Press.

Rowlands, Ian. 2000. Beauty and the Beast? BP's and Exxon's Position on Global Climate Change. *Environment and Planning C: Government and Policy* 18 (3): 339–354.

Ruitenbeek, Jack, and Cynthia Cartier. 1998. *Rational Exploitations: Economic Criteria and Indicators for Sustainable Management of Tropical Forests*. Occasional Paper No. 17, November. Bogor, Indonesia: Center for International Forestry Research.

Runge, C. Ford, and Benjamin Senauer. 2000. A Removable Feast. *Foreign Affairs* 79 (3): 39–51.

Rutherford, Paul. 2003. "Talking the Talk": Business Discourse at the World Summit on Sustainable Development. *Environmental Politics* 12 (2): 145–150.

Ryan, John, and Alan Durning. 1997. *Stuff: The Secret Lives of Everyday Things*. Seattle: Northwest Environment Watch.

Sachs, Wolfgang. 1999. *Planet Dialectics: Explorations in Environment and Development*. London: Zed Books.

Sachs, Wolfgang, ed. 1993. *Global Ecology: A New Arena of Political Conflict*. London: Zed Books.

Schafer, Kristin S. 2002. Ratifying Global Toxics Treaties: The United States Must Provide Leadership. *SAIS Review* 22 (1): 169–176.

Scherr, Sara. 2000. A Downward Spiral? Research Evidence on the Relationship between Poverty and Natural Resource Degradation. *Food Policy* 25 (1): 479–498.

Schmidheiny, Stephan (with the World Business Council for Sustainable Development). 1992. *Changing Course: A Global Business Perspective on Development and the Environment*. Cambridge, MA: MIT Press.

Schmidheiny, Stephan, and Federico Zorraquin. 1996. *Financing Change: The Financial Community, Eco-Efficiency and Sustainable Development*. Cambridge, MA: MIT Press.

Scholte, Jan Aart. 1997. The Globalization of World Politics. In John Baylis and Steve Smith, eds., *The Globalization of World Politics: An Introduction to International Relations*, 13–30. Oxford: Oxford University Press.

Scholte, Jan Aart. 2000. *Globalization: A Critical Introduction*. New York: St. Martin's Press.

Schor, Juliet. 1998. *The Overspent American: Upscaling, Downshifting, and the New Consumer*. New York: Basic Books.

Schreurs, Miranda A. 2002. *Environmental Politics in Japan, Germany, and the United States*. Cambridge: Cambridge University Press.

Schumacher, E. F. 1973. *Small Is Beautiful: Economics as if People Mattered*. New York: Harper and Row.

Selin, Henrik, and Noelle Eckley. 2003. Science, Politics, and Persistent Organic Pollutants: Scientific Assessments and Their Role in International Environmental Negotiations. *International Environmental Agreements: Politics, Law and Economics* 3 (1): 17–42.

Selin, Henrik, and Stacy D. VanDeveer. 2003. Mapping Institutional Linkages in European Air Pollution Politics. *Global Environmental Politics* 3 (3): 14–46.

Sen, Amartya. 1981. *Poverty and Famines: An Essay on Entitlement and Deprivation*. Oxford: Oxford University Press.

Sen, Gita. 1994. Women, Poverty and Population: Issues for the Concerned Environmentalist. In Wendy Harcourt, ed., *Feminist Perspectives on Sustainable Development*, 215–225. London: Zed Books.

Shiva, Mira. 1994. Environmental Degradation and Subversion of Health. In Vandana Shiva, ed., *Close to Home: Women Reconnect Ecology, Health and Development Worldwide*, 60–77. Gabriola Island, BC: New Society.

Shiva, Vandana. 1989. *Staying Alive: Women, Ecology and Development*. London: Zed Books.

Shiva, Vandana. 1992. *The Violence of the Green Revolution*. London: Zed Books.

Shiva, Vandana. 1993a. GATT, Agriculture and Third World Women. In Maria Mies and Vandana Shiva, eds., *Ecofeminism*, 231–245. London: Zed Books.

Shiva, Vandana. 1993b. The Greening of the Global Reach. In Wolfgang Sachs, ed., *Global Ecology: A New Arena of Political Conflict*, 149–156. London: Zed Books.

Shiva, Vandana. 1993c. *Monocultures of the Mind: Perspectives on Biodiversity and Biotechnology*. London: Zed Books.

Shiva, Vandana. 1997. *Biopiracy: The Plunder of Nature and Knowledge*. Toronto: Between the Lines.

Shiva, Vandana. 2000. *Stolen Harvest: The Hijacking of the Global Food Supply*. Cambridge, MA: South End Press.

Simon, Julian L. 1981. *The Ultimate Resource*. Princeton, NJ: Princeton University Press.

Simon, Julian L. 1990. *Population Matters*. New Brunswick, NJ: Transaction.

Simon, Julian L. 1996. *The Ultimate Resource 2*. Princeton, NJ: Princeton University Press.

Singer, Peter. 1977. Famine, Affluence, and Morality. In William Aiken and Hugh LaFollette, eds., *World Hunger and Moral Obligation*, 572–581. Englewood Cliffs, NJ: Prentice Hall.

Sizer, Nigel. 1995. *Backs to the Wall in Surinam: Forest Policy in a Country in Crisis*. Washington, DC: World Resources Institute.

Skjærseth, Jon Birger, and Tora Skodvin. 2001. Climate Change and the Oil Industry: Common Problems, Different Strategies. *Global Environmental Politics* 1 (3): 18–42.

Sklair, Leslie. 1993. *Assembling for Development*. San Diego: Center for U.S.–Mexican Studies, University of California.

Sklair, Leslie. 2001. *The Transnational Capitalist Class*. London: Blackwell.

Smith, Adam. 1776. *An Inquiry into the Nature and Causes of the Wealth of Nations*. Available at www.adamsmith.org/smith/won-index.htm.

Soroos, Marvin S. 1997. *The Endangered Atmosphere*. Columbia: University of South Carolina Press.

Soroos, Marvin S. 2001. Global Climate Change and the Futility of the Kyoto Process. *Global Environmental Politics* 1 (2): 1–9.

Speer, Lawrence. 1997. From Command-and-Control to Self-Regulation: The Role of Environmental Management Systems. *International Environment Reporter* 20 (5): 227–228.

Steinberg, Paul F. 2001. *Environmental Leadership in Developing Countries: Transnational Relations and Biodiversity Policy in Costa Rica and Bolivia.* Cambridge, MA: MIT Press.

Stern, David, Michael Common, and Edward Barbier. 1996. Economic Growth and Environmental Degradation: The Environmental Kuznets Curve and Sustainable Development. *World Development* 24 (7): 1151–1160.

Stevis, Dimitris, and Valerie Assetto, eds. 2001. *The International Political Economy of the Environment: Critical Perspectives.* Boulder, CO: Lynne Reinner.

Stevis, Dimitris, and Stephen Mumme. 2000. Rules and Politics in International Integration: Environmental Regulation in NAFTA and the EU. *Environmental Politics* 9 (4): 20–42.

Stilwell, Matthew, and Richard Tarasofsky. 2001. *Towards Coherent Environmental and Economic Governance: Legal and Practical Approaches to MEA-WTO Linkages.* Gland, Switzerland: WWF Network. Available at www.ciel.org/Publications/Coherent_EnvirEco_Governance.pdf.

Stoett, Peter J. 1997. *The International Politics of Whaling.* Vancouver: UBC Press.

Stokke, Olav Schram, and Øystein B. Thommessen, eds. 2003. *Yearbook of International Co-operation on Environment and Development.* London: Earthscan. Available at www.greenyearbook.org.

Streck, Charlotte. 2001. The Global Environment Facility—A Role Model for International Governance? *Global Environmental Politics* 1 (2): 71–94.

Strong, Maurice. 2000. *Where on Earth Are We Going?* Toronto: Knopf.

Suri, Vivek, and Duane Chapman. 1998. Economic Growth, Trade and Energy: Implications for the Environmental Kuznets Curve. *Ecological Economics* 25 (2): 195–208.

Susskind, Lawrence. 1992. New Corporate Roles in Global Environmental Treaty-Making. *Columbia Journal of World Business* 27 (3/4): 62–73.

Susskind, Lawrence. 1994. *Environmental Diplomacy: Negotiating More Effective Global Agreements.* Oxford: Oxford University Press.

Switzer, Jacqueline Vaughn. 2004. *Environmental Politics: Domestic and Global Dimensions.* Belmont, CA: Thompson/Wadsworth.

Tabb, William. 2000. After Seattle: Understanding the Politics of Globalization. *Monthly Review* 51 (10): 1–18.

Tamiotti, Ludivine, and Matthias Finger. 2001. Environmental Organizations: Changing Roles and Functions in Global Politics. *Global Environmental Politics* 1 (1): 56–76.

Taylor, Bron Raymond, ed. 1995. *Ecological Resistance Movements: The Global Emergence of Radical and Popular Environmentalism.* Albany, NY: SUNY/Press.

Thomas, Chris, Alison Cameron, Rhys Green, Michel Bakkenes, Linda Beaumont, Yvonne C. Collingham, Barend Erasmus, Marinez Ferriera de Siqueira, Alan Grainger, Lee Hannah, Lesley Hughes, Brian Huntley, Albert Van Jaarsveld, Guy Midgley, Lera Miles, Miguel Ortega-Huerta, A. Townsend Peterson, Oliver Phillips, and Stephen Williams. 2004. Extinction Risk from Climate Change. *Nature* 427 (January 8): 145–148.

Thomas, Vinod, and Tamara Belt. 1997. Growth and the Environment: Allies or Foes? *Finance and Development* (June): 22–24.

Thompson, Peter, and Laura A. Strohm. 1996. Trade and Environmental Quality: A Review of the Evidence. *Journal of Environment and Development* 5 (4): 365–390.

Tidsell, Clem. 1993. *Environmental Economics. Policies for Environmental Management and Sustainable Development.* Cheltenham, UK: Edward Elgar.

Tidsell, Clem. 2001. Globalization and Sustainability: Environmental Kuznets Curve and the WTO. *Ecological Economics* 39 (2): 185–196.

Tierney, John. 1990. Betting the Planet. *New York Times Magazine* (December 2): 52–81.

Tietenberg, Thomas. 1992. *Environmental and Natural Resource Economics.* 3rd ed. New York: HarperCollins.

Tokar, Brian. 1997. *Earth for Sale: Reclaiming Ecology in the Age of Corporate Greenwash.* Boston: South End Press.

Tolba, Mostafa K., and Iwona Rummel-Bulska. 1998. *Global Environmental Diplomacy: Negotiating Environmental Agreements for the World 1973–1992.* Cambridge, MA: MIT Press.

Toye, John. 1991. Ghana. In Paul Mosley, Jane Harrigan, and John Toye, eds., *Aid and Power*, vol. 2, *Case Studies*, 151–200. London: Routledge.

Turaga, Uday. 2000. Damming Waters and Wisdom: Protest in the Narmada River Valley. *Technology in Society* 22: 237–253.

Union of International Associations. 1999. *Yearbook of International Organizations 1998–1999.* Brussels: Union of International Associations. Available at www.uia.org.

United Nations (UN). 1945. *Charter of the United Nations.* Available at www.unhchr.ch/pdf/UNcharter.pdf.

United Nations (UN). 1992. *Agenda 21: The United Nations Programme of Action from Rio.* New York: United Nations.

United Nations (UN). 2002. *Johannesburg Declaration on Sustainable Development.* Available at www.johannesburgsummit.org/html/documents/summit_docs/1009wssd_pol_declaration.doc.

United Nations (UN). 2003. *UNDP Budget Estimates for the Biennium 2004–2005*, DP/2003/28. Available at www.undp.org/execbrd/pdf/dp03-28e.pdf.

United Nations (UN). No date. *Rules of the Governing Council.* Available at www.unep.org/Documents/Default.asp?DocumentID=77&ArticleID=1157.

United Nations Conference on Trade and Development (UNCTAD). 2000. *The Least Developed Countries 2000 Report.* Geneva: United Nations Conference on Trade and Development.

United Nations Conference on Trade and Development (UNCTAD). 2001. *World Investment Report 2001: Promoting Linkages.* Available at www.unctad.org/en/docs/wir01ove_a4.en.pdf.

United Nations Conference on Trade and Development (UNCTAD). 2002. *World Investment Report 2002: Transnational Corporations and Export Competitiveness.* New York: United Nations. (Also used data from the 1990, 1995, and 2000 *Reports.*)

United Nations Conference on Trade and Development (UNCTAD), Program on Transnational Corporations. 1993. *Environmental Management in Transnational Corporations: Report on the Benchmark Corporate Environmental Survey.* Environment Series No. 4. New York: United Nations.

United Nations Development Programme (UNDP). 1998. *Human Development Report 1998: Consumption for Human Development.* Oxford: Oxford University Press.

United Nations Development Programme (UNDP). 1999. *Human Development Report 1999: Globalization with a Human Face.* Oxford: Oxford University Press.

United Nations Development Programme (UNDP). 2001. *Human Development Report 2001: Making New Technologies Work for Human Development.* Oxford: Oxford University Press.

United Nations Development Programme (UNDP). 2003. *Human Development Report 2003: Millennium Development Goals: A Compact Among Nations to End Human Poverty.* Oxford: Oxford University Press.

United Nations Development Programme (UNDP) and United Nations Fund for Population Activities (UNFPA). 2002. *Executive Board of UNDP and UNFPA.* Available at www.undp.org/execbrd/pdf/eb-overview.PDF.

United Nations Environment Programme (UNEP). 2000. *Action on Ozone.* Available at www.unep.ch/ozone/pdf/ozone-action-en.pdf.

United Nations Environment Programme (UNEP). 2001a. *International Environmental Governance: Report of the Executive Director to the Open-Ended Intergovernmental Group of Ministers or Their Representatives on International Environmental Governance.* First meeting: New York, April 18, 2001. Available at www.unep.org/IEG/docs/working%20documents/reportfromED/IGM_1_2.E.doc.

United Nations Environment Programme (UNEP). 2001b. News Release 01/40; *World Water Day 2001: Water for Health.* Available at www.unep.org/Documents/Default.asp?DocumentID=193&ArticleID=2801.

United Nations Environment Programme (UNEP). 2002a. *Capacity Building for Sustainable Development: An Overview of UNEP Environmental Capacity Development Activities.* Geneva: United Nations Environment Programme.

United Nations Environment Programme (UNEP). 2002b. *Global Environment Outlook 3*. London: Earthscan.

United Nations Environment Programme (UNEP). No date. *UNEP Resource Mobilization*. Available at www.unep.org/rmu/.

United Nations Fund for Population Activities (UNFPA). 2001. *The State of the World Population 2001. Footprints and Milestones: Population and Environmental Change*. New York: United Nations Fund for Population Activities. Available at www.unfpa.org/swp/2001/english/index.html.

United Nations Transnational Corporations and Management Division, Department of Economic and Social Development. 1992. *World Investment Report 1992*. New York: United Nations.

VanDeveer, Stacy D., and Geoffrey D. Dabelko. 2001. It's Capacity, Stupid: International Assistance and National Implementation. *Global Environmental Politics* 1 (2): 18–29.

Victor, David, Kal Raustiala, and Eugene Skolnikoff, eds. 1998. *The Implementation and Effectiveness of International Environmental Commitments: Theory and Practice*. Cambridge, MA: MIT Press.

Vogel, David. 2000. International Trade and Environmental Regulation. In Norman J. Vig and Michael E. Kraft, eds., *Environmental Policy: New Directions for the Twenty-First Century*, 4th ed., 371–389. Washington, DC: CQ Press.

Vogler, John. 2000. *The Global Commons: Environmental and Technological Governance*. 2nd ed. New York: Wiley.

Vogler, John. 2003. Taking Institutions Seriously: How Regime Analysis Can Be Relevant to Multilevel Environmental Governance. In David Humphreys, Matthew Paterson, and Lloyd Pettiford, eds., *Global Environmental Governance for the Twenty-First Century: Theoretical Approaches and Normative Considerations*. Special issue of *Global Environmental Politics* 3 (2): 25–39.

von Moltke, Konrad. 2001. The Organization of the Impossible. *Global Environmental Politics* 1 (1): 23–28.

Wackernagel, Mathis, and William Rees. 1996. *Our Ecological Footprint: Reducing Human Impact on the Earth*. Gabriola Island, BC: New Society.

Wackernagel, Mathis, Niels B. Schulz, Diana Deumling, Alejandro Callejas Linares, Martin Jenkins, Valerie Kapos, Chad Monfreda, Jonathan Loh, Norman Myers, Richard Norgaard, and Jørgen Randers. 2002. Tracking the Ecological Overshoot of the Human Economy. *Proceedings of the National Acadamy of Sciences* 99 (14): 9266–9271.

Wade, Robert. 1997. Greening the Bank: The Struggle over the Environment, 1970–1995. In John Lewis and Richard Webb, eds., *The World Bank: Its First Half Century*, vol. 2, 611–735. Washington, DC: Brookings Institution Press.

Wade, Robert. 2001. Global Inequality: Winners and Losers. *The Economist* (April 28): 72–74.

Wallach, Lori, and Michelle Sforza. 1999. *Whose Trade Organization? Corporate Globalization and the Erosion of Democracy.* Washington, DC: Public Citizen.

Wapner, Paul. 1996. *Environmental Activism and World Civic Politics.* Albany, NY: SUNY Press.

Wapner, Paul. 2002. Horizontal Politics: Transnational Environmental Activism and Global Cultural Change. *Global Environmental Politics* 2 (2): 37–62.

Wapner, Paul. 2003. World Summit on Sustainable Development: Toward a Post-Jo'burg Environmentalism. *Global Environmental Politics* 3 (1): 1–10.

Waring, Marilyn. 1999. *Counting for Nothing: What Men Value and What Women Are Worth.* Toronto: University of Toronto Press.

Watts, Phil, and Richard Holme. 1999. *Corporate Social Responsibility: Meeting Changing Expectations.* Geneva: World Business Council for Sustainable Development.

Weinstein, Michael, and Steve Charnovitz. 2001. The Greening of the WTO. *Foreign Affairs* 80 (6): 147–156.

Weiss, Edith Brown, and Harold K. Jacobson, eds. 1998. *Engaging Countries: Strengthening Compliance with International Environmental Accords.* Cambridge, MA: MIT Press.

Weiss, Linda. 1998. *The Myth of the Powerless State.* Ithaca, NY: Cornell University Press.

Welford, Richard. 1997. *Hijacking Environmentalism: Corporate Responses to Sustainable Development.* London: Earthscan.

Wenz, Peter S. 2001. *Environmental Ethics Today.* Oxford: Oxford University Press.

Wettestad, Jörgen. 1999. *Designing Effective Environmental Regimes: The Key Conditions.* Cheltenham, UK: Edward Elgar.

Whalley, John, and Ben Zissimos. 2001. What Could a World Environmental Organization Do? *Global Environmental Politics* 1 (1): 29–34.

Wheeler, David. 2001. Racing to the Bottom? Foreign Investment and Air Pollution in Developing Countries. *Journal of Environment and Development* 10 (3): 225–245.

Wheeler, David. 2002. Beyond Pollution Havens. *Global Environmental Politics* 2 (2): 1–10.

Willetts, Peter. 2001. Transnational Actors and International Organizations in Global Politics. In John Bayliss and Steve Smith, eds., *The Globalization of World Politics: An Introduction to International Relations,* 356–383. Oxford: Oxford University Press.

Williams, Edward. 1996. The Maquiladora Industry and Environmental Degradation in the United States–Mexico Borderlands. *St. Mary's Law Journal* 27 (4): 777–779.

Williams, Marc. 1994. International Trade and the Environment: Issues, Perspectives, and Challenges. In Caroline Thomas, ed., *Rio: Unravelling the Consequences*, 80–97. Ilford, UK: Frank Cass.

Williams, Marc. 2001. In Search of Global Standards: The Political Economy of Trade and the Environment. In Dimitris Stevis and Valerie Assetto, eds., *The International Political Economy of the Environment: Critical Perspectives*, 39–61. Boulder, CO: Lynne Reinner,

Wilson, Edward O. 2002. *The Future of Life*. New York: Vintage Books.

World Bank. 1992. *World Development Report 1992*. New York: Oxford University Press.

World Bank. 1994. *Adjustment in Africa: Reforms, Results and the Road Ahead*. Oxford: Oxford University Press.

World Bank. 1995. *Mainstreaming the Environment: The World Bank Group and the Environment Since the Rio Earth Summit*. Washington, DC: World Bank.

World Bank. 1999. *Greening Industry: New Roles for Communities, Markets and Governments*. Oxford: Oxford University Press. Available at www.worldbank/org/nipr/greening.

World Bank. 2001a. *Adjustment Lending Retrospective: Final Report, Operations Policy and Country Services*. Available at lnweb18.worldbank.org/ext/language.nsf/0a70c4735d4fc79e852566390063b35c/eec155910a2720a885256 ae30076a40c/$FILE/alretro.pdf.

World Bank. 2001b. *Making Sustainable Commitments: An Environment Strategy for the World Bank*. Washington, DC: World Bank. Available at www.worldbank.org/environment/index.htm.

World Bank. 2003a. *Global Development Finance*. Washington, DC: World Bank.

World Bank. 2003b. *World Development Report 2003. Sustainable Development in a Dynamic World: Transforming Institutions, Growth, and Quality of Life*. Washington, DC: World Bank.

World Business Council for Sustainable Development (WBCSD). 2000. *Building a Better Future: Innovation, Technology and Sustainable Development*. Available at www.wbcsd.org/DocRoot/xPV0RGh50wmrkQHs1oik/Building.pdf.

World Business Council for Sustainable Development (WBCSD). 2002. *The Business Case for Sustainable Development: Making a Difference toward the Johannesburg Summit 2002 and Beyond*. Geneva: World Business Council for Sustainable Development.

World Commission on Environment and Development (WCED). 1987. *Our Common Future*. Oxford: Oxford University Press.

World Health Organization (WHO). 2002. *Swedish Conference Seeks to Reduce Mortality of Babies and Young Children*. March 15. Available at www.europaworld.org/issue73/swedishconference15302.htm.

World Meteorological Organization (WMO). 1997. *Comprehensive Assessment of the Freshwater Resources of the World*. Geneva: WMO.

World Resource Institute (WRI). 1997. *The Last Frontier Forests: Ecosystems and Economies on the Edge.* Washington, DC: WRI.

World Trade Organization (WTO). 1998a. *Understanding the WTO.* Available at www.wto.org/english/thewto_e/whatis_e/tif_e/tif_e.htm.

World Trade Organization (WTO). 1998b. *The WTO, The Uruguay Round: Module 1 FAQs.* Available at www.wto.org/english/thewto_e/whatis_e/eol/e/default.htm.

World Trade Organization (WTO). 1999. WTO, Trade and Environment. Special Studies 4. Geneva: World Trade Organization.

World Trade Organization (WTO). 2001. International Trade Statistics 2001. Available at www.wto.org/english/res_e/statis_e/its2001_e/chp_2_e.pdf.

World Trade Organization (WTO). 2003. *WTO Secretariat Budget for 2003.* Available at www.wto.org/english/thewto_e/secre_e/budget03_e.htm.

Worldwatch Institute. 2003. *State of the World 2003.* Washington, DC: Worldwatch. Summary at www.worldwatch.org/press/news/2003/01/09.

WWF Network. 2002. *Living Planet Report 2002.* Gland, Switzerland: WWF–World Wide Fund for Nature. Available at www.panda.org/downloads/general/LPR_2002.pdf.

Yoder, Andrew J. 2003. Lessons from Stockholm: Evaluating the Global Convention on Persistent Organic Pollutants. *Indiana Journal of Global Legal Studies* 10 (Summer): 113–156.

Young, Oran R. 1989. *International Cooperation: Building Regimes for Natural Resources and the Environment.* Ithaca, NY: Cornell University Press.

Young, Oran R. 1994. *International Governance: Protecting the Environment in a Stateless Society.* Ithaca, NY: Cornell University Press.

Young, Oran R. 2001. Evaluating the Effectiveness of International Environmental Regimes. *Global Environmental Politics* 1 (1): 99–121.

Young, Oran R. 2002. *The Institutional Dimensions of Environmental Change: Fit, Interplay, and Scale.* Cambridge, MA: MIT Press.

Young, Oran R., ed. 1999. *The Effectiveness of International Environmental Regimes: Causal Connections and Behavioral Mechanisms.* Cambridge, MA: MIT Press.

Young, Zoe. 2003. *A New Green Order: The World Bank and the Politics of the Global Environment Facility.* London: Pluto Press.

Zarsky, Lyuba. 2002. APEC: The "Sustainable Development" Agenda. In Richard Steinberg, ed., *The Greening of Trade Law: International Trade Organizations and Environmental Issues,* 221–247. Lanham, MD: Rowman and Littlefield.

Zuckerman, Ben. 2004. Nothing Racist about It. *Globe and Mail* (January 28): A17.

Index